M000207646

Failed Promises

American and Comparative Environmental Policy

Sheldon Kamieniecki and Michael E. Kraft, series editors

For a complete list of books in the series, please see the back of the book.

Failed Promises

Evaluating the Federal Government's Response to Environmental Justice

Edited by David M. Konisky

The MIT Press
Cambridge, Massachusetts
London, England

© 2015 Massachusetts Institute of Technology

All rights reserved. No part of this book may be reproduced in any form by any electronic or mechanical means (including photocopying, recording, or information storage and retrieval) without permission in writing from the publisher.

MIT Press books may be purchased at special quantity discounts for business or sales promotional use. For information, please email special_sales@mitpress.mit.edu.

This book was set in Sabon by the MIT Press. Printed and bound in the United States of America.

Library of Congress Cataloging-in-Publication Data is available.

ISBN: 978-0-262-02883-7 (hardcover); 978-0-262-52735-4 (paperback)

10 9 8 7 6 5 4 3 2 1

To my parents

Contents

Series Foreword

Environmental justice issues rose to prominence in the 1980s and 1990s as evidence mounted that some communities were more likely than others to be exposed to health risks associated with toxic chemicals and hazardous waste. At that time, scholarly interest in environmental equity issues was relatively new, and the early literature grew out of case studies of communities that seemed to be targeted for disposal of hazardous wastes and other environmental risks. A lively scholarly debate ensued as the literature matured, with new research on how best to define and measure environmental inequities across the nation, why they occurred, and what might be done about them. Many of the studies in this round of scholarship employed rigorous empirical methodologies, and sometimes the researchers reached conclusions that challenged earlier findings on the extent to which environmental inequity could be confirmed, the causes of it, and whether public health was threatened.

In the United States, environmental justice usually refers to the disproportional exposure to pollution and its health risks or the unequal receipt of benefits related to the implementation of environmental policies that are designed to reduce health risks, for example, the Clean Air Act for urban air pollution and the Resource Conservation and Recovery Act for the handling of hazardous waste. As the contributors to this volume describe so well, concern has focused on the disproportionate environmental and health risks experienced by low-income and racial and ethnic minorities, particularly African-Americans, Hispanics, and Native Americans. These risks come from many sources, among the most prominent of which are urban air pollution, hazardous and solid waste landfills and incinerators, and uranium mining on Western lands.

In many respects, as the authors tell us, the politics of environmental justice differ from what we might consider typical politics of the environment, where scholars analyze governmental institutions, decision making within them, and the policies that emerge. Historically, the politics of environmental justice actions instead has involved the activities of grassroots organizations in minority or low-income communities with a history of unusual environmental exposures that they believed were either ignored by policymakers or given insufficient consideration. Not surprisingly, much early work on environmental justice focused on these communities and the activists within them who sought to bring about change.

Much of this pattern of political activity, however, shifted during the 1990s, particularly during the Clinton administration. President Clinton's team offered new policy mandates, administrative reorganization, and research and data collection programs. The Clinton team also mounted efforts to reach out to communities and build their capacity to respond to environmental risks. The president's broad Executive Order 12898, signed in 1994, in many ways was the most dramatic change of all. It called on all federal agencies to incorporate equity considerations into their programs, policies, and activities. As this book puts it squarely, the promotion of environmental equity was to become a "core consideration in federal environmental decision making." It is this new set of goals and political activity to which the contributors to this volume direct their attention. They seek to discover whether the aspirations embodied in the executive order and related actions have been realized. What has been accomplished as a result of the new level of political attention given to environmental inequity and the new policy priorities?

With this focus, the authors address questions that were left unanswered by previous work on environmental justice that concentrated on documenting the extent of environmental inequity and its causes, as well as examining the activities of community activists. The contributors to this book shift attention to the processes of policy implementation and policy change following the Clinton executive order. They want to know how well the policies have been implemented by executive agencies and what results have followed.

At the center of this new inquiry are the impacts of policy change. Did the new policies and actions make a difference, and, if so, how much of a difference? And what factors made the policy actions a success? The flip side of the question is equally important. If policy changes did not work as expected, why is that? What factors constrained policy implementation or limited the achievements? If the real world impacts of environmental justice policies, or other comparable policy actions, are minor and largely symbolic, as the contributors argue is the case, it is important to document that and to explain why the policies did not achieve more so that policy analysts can understand this outcome. The public and policymakers may choose to mount additional policy efforts with knowledge of why previous policy efforts failed.

This book was written on the twentieth anniversary of President Clinton's executive order, and it takes what the authors call a "hard look" at the impact of that order, particularly within the EPA. The contributors come to this task with a diversity of disciplinary backgrounds suitable for analyzing the different dimensions of the EPA's performance and reviewing previous research on environmental justice. As David Konisky observes, the chapters in this book constitute the first comprehensive evaluation of federal environmental justice policy as it has been implemented over the past twenty years. The chapters are organized around three widely-recognized aspects of environmental justice: distributive justice, procedural justice, and corrective justice. While the conclusion is that the federal government has thus far failed to live up to the promise of the executive order, the contributors also find a new concern for environmental justice at the EPA, with significant steps being taken to translate the initial concerns into specific changes in rulemaking, permitting, enforcement, and community engagement.

The book illustrates well our purpose in the MIT Press series in American and Comparative Environmental Policy. We encourage work that examines a broad range of environmental policy issues. We are particularly interested in volumes that incorporate interdisciplinary research and focus on the linkages between public policy and environmental problems and issues both within the United States and in cross-national settings. We welcome contributions that analyze the policy dimensions of relationships between humans and the environment from either a theoretical or empirical perspective.

At a time when environmental policies are increasingly seen as controversial and new and alternative approaches are being implemented widely, we especially encourage studies that assess policy successes and failures, evaluate new institutional arrangements and policy tools, and clarify new directions for environmental politics and policy. The books in this series are written for a wide audience that includes academics, policymakers, environmental scientists and professionals, business and labor leaders, environmental activists, and students concerned with environmental issues. We hope they contribute to public understanding of environmental problems, issues, and policies of concern today and also suggest promising actions for the future.

Sheldon Kamieniecki, University of California, Santa Cruz
and
Michael Kraft, University of Wisconsin-Green Bay
Co-editors, American and Comparative Environmental Policy Series

Preface

The United States has made a lot of progress in protecting the environment. Forty years ago, a strong bipartisan coalition in Congress enacted a suite of federal pollution control laws to reduce air and water pollution, improve the safety of drinking water, safeguard the disposal of solid and hazardous waste, and clean up contaminated sites. Congress was responding to strong citizen demand for action, spurred on by high profile events such as the oil spill off the coast of Santa Barbara, the Cuyahoga River famously catching fire, suffocating air pollution in many major cities, and toxic waste scares in places like Niagara Falls, New York. As a result of these federal efforts, as well as the initiative of many state and local governments, most Americans enjoy a better quality of life.

Important environmental challenges certainly remain. Congress has failed to effectively address critical problems such as climate change and nonpoint source water pollution. And, many of the nation's core pollution control laws are in desperate need of reform to improve their economic efficiency, uneven implementation, and capacity to address new risks to human and ecological health. There is little doubt, however, that our air is safer to breathe, our water is safer to drink, and our food is safer to eat.

Yet, the dividends of these significant efforts to improve environmental quality have not been evenly distributed across the country. We know from decades of scholarly research that some segments of the U.S. population, particularly poor and minority communities, experience disproportionate environmental burdens and enjoy fewer environmental amenities. Concerns about these race- and class-based disparities spawned a new, grassroots social movement dedicated to achieving "environmental justice" for these communities. The environmental justice movement came to life in the mid-1980s, first to fight against locally unwanted land uses, and over time, to forcefully demand fairness in environmental outcomes and to gain equal access to government resources and decision making.

After nearly a decade of political organization and advocacy, the environmental justice movement successfully garnered the attention of the federal government in the early 1990s. The Clinton Administration initiated a series of administrative actions that formally brought environmental justice onto the agenda of the U.S. Environmental Protection Agency (EPA) and other federal agencies. These actions included new policy mandates and strategies, administrative reorganization, new research and data collection programs, enhanced outreach and community-capacity building programs, and, most notably, a presidential executive order on environmental justice. Executive Order 12898 (Federal Actions to Address Environmental Justice in Minority and Low-Income Populations) (EO 12898), signed by President Bill Clinton on February 11, 1994, called on each federal agency to "make achieving environmental justice part of its mission by identifying and addressing, as appropriate, disproportionately high and adverse human health or environmental effects of its programs, policies, and activities on minority populations and low-income populations." In essence, EO 12898 mandated that federal agencies devise an environmental justice strategy, consider the distributional impacts of their actions and policies, and open up their decision making to a more diverse group of stakeholders, especially members of minority and lower-income communities.

Given its central role in administering the nation's major environmental statutes, the EPA was tasked with much of the responsibility for implementing these new policy mandates. The executive order required the EPA to coordinate environmental justice efforts throughout the federal bureaucracy, and, more importantly, to integrate equity considerations throughout the agency's core regulatory activities and programs. In the twenty years since President Clinton signed EO 12898, a number of studies have analyzed specific dimensions of federal environmental justice policy, but, to date, there has not been a systematic evaluation of the implementation of the executive order or the other environmental justice policy commitments made by the EPA in the years that followed. The principal objective of this book is to provide such an evaluation.

The chapters in this book carefully examine federal environmental justice policy as it has been carried out over the past two decades, with an emphasis on the performance of the EPA in implementing EO 12898. Each of the contributing authors focuses on a different area of environmental decision making, including permitting, standard-setting, economic analysis, public participation, enforcement, and use of the courts. The authors employ different methodologies and use different types of evidence, but the analyses reach a similar general conclusion: the federal government, and the EPA in particular, has generally failed to deliver on the promises articulated in EO 12898 and in subsequent policy commitments.

Despite this discouraging conclusion, the authors also provide reason for optimism. Former EPA Administrator Lisa Jackson prioritized environmental justice during her tenure, and as a result it has regained prominence on the agency's agenda. For the first time in two decades, the EPA is taking significant steps through its development of *Plan EJ 2014* to translate the intent of EO 12898 into specific policies and procedures. Only time will tell if this new policy initiative generates significant and sustained changes in federal environmental decision making. To be sure, environmental justice advocates will be watching the performance of the EPA and the rest of the federal government. And, for our part, we in the academic community should be poised to evaluate whether these new initiatives achieve better results than policy efforts to date.

David M. Konisky
Washington, D.C.
September 2014

Acknowledgments

This book is a result of the efforts of a great many people. Let me first recognize the outstanding group of scholars that contributed chapters, Dorothy Daley, Eileen Gauna, Elizabeth Gross, Doug Noonan, Tony Reames, Chris Reenock, Ron Shadbegian, Paul Stretesky, and Ann Wolverton. Each of them brought extraordinary expertise and thoughtfulness to the project. They were also a delight to work with, and I sincerely appreciate their responsiveness to my editorial requests and sometimes tight deadlines. The book also benefitted enormously from the feedback and suggestions of three anonymous reviewers, and conversations I had with a number of people over the past few years, including Laura Anderko, Bill Gormley, Paul Mohai, Evan Ringquist, and Shalini Vajjhala.

I would like to express particular gratitude to Evan Ringquist. Evan's research, including his important work on environmental justice, continues to inspire me to pursue difficult questions in environmental politics and policy. I am indebted to his kindness and mentorship to me as a young scholar. His recent passing is a great loss to us all.

I have been fortunate to be supported by smart, motivated, and capable research assistants and I have worked with none better than Carly Morrison, Tyler Schario, and Owen Witek, each of whom assisted with this project at different stages.

Mike Kraft and Sheldon Kamieniecki enthusiastically supported the project from the outset, and I am very pleased to have the book included as part of their series at MIT Press. Clay Morgan was also a strong supporter of the book, and I appreciate his efforts to efficiently marshal the book through the review process. Beth Clevenger, Miranda Martin, Marcy Ross, and the entire production staff at MIT Press were also a joy to work with.

The book is dedicated to my parents, Jordan and Judy, whose support and encouragement is constant. My brother Daniel always provides a critical perspective, challenging my basic ideas and assumptions. Finally, I wish to thank Kristen, Benjamin, and William. Their love, support, and patience are both unconditional and endless, and they make this, and all of my work, possible.

1

Introduction

David M. Konisky

We now believe that people of color and low income are disproportionately affected by some environmental risks—the risk of living near landfills, municipal waste combustors, or hazardous waste sites. ... I have made environmental justice one of the key policy themes of my administration. Environmental justice must be woven into all aspects of EPA operations: rulemaking, permitting, enforcement, education, hiring, and outreach. (Browner 1993)
—Carol Browner testimony before U.S. House Committee on Government Operations, May 6, 1993

We have begun a new era of outreach and protection for communities historically underrepresented in EPA decision making. We are building strong working relationships with tribes, communities of color, economically distressed cities and towns, young people, and others, but this is just a start. We must include environmental justice principles in all of our decisions. This is an area that calls for innovation and bold thinking, and I am challenging all of our employees to bring vision and creativity to our programs. (Jackson 2010)
—Lisa Jackson, Memorandum to All EPA Employees, January 12, 2010

Since the enactment of the modern U.S. environmental protection system in the 1970s and early 1980s, environmental policy issues have been contested largely on the same political ground. On one side are those—led by environmental and public health advocacy organizations—that devote their energy to pushing for more stringent protections, be they to prevent pollution that causes adverse health outcomes or to protect endangered species. The other side—led by many in the business community—typically counters with the argument that such protections are too costly and therefore harmful to business competitiveness and the U.S. economy overall.[1] The major political parties now increasingly align themselves with one of these sides. With some exceptions, Democrats tend to endorse existing environmental protections and often call for tightening them, while Republicans tend to demand either the curtailment of existing policies or that, at minimum, new protections not be put into place.[2] These are the "usual politics" of contemporary U.S. environmental policy, and we see them play out repeatedly in the context of issues ranging from air pollution and chemicals to public lands and climate change.

Federal agencies are often caught in the middle, trying to balance these competing interests while simultaneously pursuing their missions. This is particularly true of the U.S. Environmental Protection Agency (EPA), which is tasked by Congress to implement a suite of environmental laws, such as the Clean Air Act (CAA), the Clean Water Act (CWA), and the Resource Conservation and Recovery Act (RCRA). These statutes aim to protect human health and the environment, often through inflexible, one-size-fits-all solutions, but the EPA is under enormous pressure to implement them without imposing overly burdensome regulations and high compliance costs on businesses large and small. This is no easy task. Further complicating matters for the EPA has been the emergence of new problems that do not fit easily into the command-and-control, technology-based infrastructure that epitomizes modern environmental laws. Environmental justice is one such problem.

Environmental justice, also referred to as *environmental equity*,[3] refers to a broad set of issues related to the distribution of environmental risks and benefits across geographic and temporal space. In the U.S. context, environmental justice most often refers to the disproportionate environmental risks experienced by low-income and racial and ethnic minorities (particularly African Americans, Hispanics, and Native Americans). Often pointed to as examples are the high number of hazardous and solid waste landfills and incinerators located in poor and minority neighborhoods, the toxic legacy of uranium mining on Native American (especially Navajo) lands, and poor air quality in urban centers.[4] In this sense, when most people refer to environmental justice or environmental equity, what they really mean is environmental *injustice* or environmental *inequity*.

The EPA defines *environmental justice* as "the fair treatment and meaningful involvement of all people regardless of race, color, national origin, or income with respect to the development, implementation, and enforcement of environmental laws, regulations, and policies" (EPA 2011). By *fair treatment*, the agency means that no segment of society should bear a disproportionate burden of environmental harms and risks, and by *meaningful involvement,* it means that individuals should have an opportunity to participate in government decisions about any activity that may result in such harms and risks to them (EPA 2011). Taken at face value, this definition does not seem particularly controversial. It simply calls for equal protection, just as do many other equal opportunity and anti-discrimination measures. But from the start, issues of environmental justice have been charged by the politics of race and class, and it is this intersection of race, class, and environmental protection that distinguishes the policy and politics of environmental justice from much of the rest of environmental policy. Environmental justice, in other words, does not conform to the "usual politics" of U.S. environmental policy.

Concerns about environmental inequality rose to the policy agendas of federal, state, and local governments largely outside of mainstream environmental politics, and some argue despite them (Bullard 1994b; Mohai, Pellow, and Timmons 2009). Environmental justice issues were brought to the attention of policymakers as the result of grassroots efforts in minority and low-income communities, first fighting against locally unwanted land use in places such as Warren County, North Carolina, Chester, Pennsylvania, and Kettleman City, California, and eventually uniting behind a common agenda that seeks fairness in environmental outcomes and equal access to decision making. The movement was motivated by a growing belief that residents of poor and minority communities suffered disproportionate environmental burdens created in large part by discriminatory practices of both private firms and public agencies. In addition, there were widely held beliefs that poor and minority interests were being systematically excluded from private-sector and government decision making, and that the burdens they experienced were exacerbated by this institutionalized bias. Moreover, many in the environmental justice movement believed that their interests were being ignored by national environmental advocacy organizations, which they accused of neglecting to represent their concerns, if not overtly discriminating against them.[5]

Beginning in the 1990s, the growing movement found a receptive audience with the incoming administration of President Bill Clinton, which embarked on a federal environmental justice initiative in response. This initiative was wide ranging and included policy mandates, administrative reorganization, new research and data collection programs, and new emphasis on outreach and community capacity building. The centerpiece of the initiative was Executive Order 12898 ("Federal Actions to Address Environmental Justice in Minority Populations and Low-Income Populations," hereafter referred to as EO 12898), which was signed by President Clinton on February 11, 1994. EO 12898 called on all agencies of the federal government to incorporate equity considerations into their programs, policies, and activities "[t]o the greatest extent practicable and permitted by law." The executive order is extremely broad in scope and intent, prioritizing consideration of equity in standard setting, permitting, and rulemaking, improved enforcement of existing statutes, greater public participation in decision making, and targeted data collection on the environmental and health risks facing minority and low-income populations. As reflected in the two quotes at the beginning of the chapter from former EPA administrators Carol Browner and Lisa Jackson, the EPA in particular has repeatedly made commitments to integrate environmental justice considerations into day-to-day decision making across a wide set of its activities.

The types of policies expressed in the federal government's environmental justice initiatives deviate from the way environmental policy programs are usually conceived. Environmental protection is typically thought of as a public good that benefits everyone. Advocates for environmental justice challenge the very public-good nature of environmental quality and threaten to undercut the broad level of public support characteristic of most pollution control efforts (Ringquist 2004). In essence, advocates transform environmental protection into more of a redistributive framework, which tends to bring with it much more conflictual politics. It also moves the EPA away from the technocratic decision making to which it is accustomed (Keuhn 2000). Thus, in contrast to most environmental regulations, which establish performance criteria or standards, the environmental justice mandates embodied in EO 12898 and elsewhere focus on achieving environmental justice as both a process and an aspirational outcome. Environmental equity was to become a core consideration in federal environmental decision making.

Has this been accomplished? Writing fifteen years ago, Terry Davies and Jan Mazurek (1998) commented that "[l]ittle systematic evaluation has been conducted to measure how well the programs and initiatives of EPA's Environmental Justice Strategy have worked to improve environmental quality for minority and low income communities" (p. 184).

The same can be said today. To be sure, many of the initiatives put forward have run into roadblocks. As Evan Ringquist (2004) has eloquently remarked, "these ambitious goals and initiatives have beached routinely on the shoals of a formidable mix of scientific, political, organizational, and financial obstacles" (p. 257). A number of studies have pointed to significant shortcomings in specific elements of the federal government's implementation of EO 12898 and other environmental justice policy mandates (National Academy of Public Administration 2001, EPA 2004; GAO 1995). To date, however, there has been neither a systematic description of the implementation process nor a rigorous review of the efficacy of federal government policy. Have the EPA and other federal agencies moved to make environmental justice a core component of their decision making in permitting, standard setting, economic analysis, enforcement, and other areas? Or, instead, has achieving environmental justice resulted in mostly symbolic actions, with few concrete steps taken to address inequities? Perhaps the reality lies somewhere in the middle.

The answers to these questions are not just academic. Extensive research shows that low-income and racial and ethnic minorities continue to experience health disparities due to environmental exposures (e.g., Adler and Newman 2002; Brulle and Pellow 2006; Clark, Millet, and Marshall 2014; Gee and Payne-Sturges 2004; Institute of Medicine 2002; Sexton 2000; EPA 2003; Yen and Syme 1999). Other studies show that these groups continue to live in closer proximity to hazardous waste sites and noxious facilities (e.g., Arora and Cason 1999; Mohai and Saha 2006, 2007; Pollock and Vittas 1995; Ringquist 1997). In other words, environmental justice is not just a historical problem—it is a contemporary one. In addition, to the extent that race- and class-based disparities occur because of failures of government agencies to fairly carry out environmental protection, as some research has shown (Helland 1998; Konisky 2009a; Konisky and Reenock 2013b; Konisky and Schario 2010; Scholz and Wang 2006), these disparities also represent a basic violation of equal protection under the law, a core principle of democratic public policy. Finally, the EPA is currently undertaking a massive "rethink" of its approach to environmental justice, designing new policies, procedures, and tools to better support achievement of equity goals. These policies will continue to be developed in the years to come, and they would certainly benefit from the lessons learned from a careful analysis of past experience.

As we mark the twenty-year anniversary of EO 12898, it is an opportune time for such an analysis. This book takes a hard look at the impact of this executive order, with particular emphasis on the EPA given the agency's central role in environmental decision making. The interdisciplinary group of contributors to this volume analyze various dimensions of EPA performance (and, at times, the performance of other relevant agencies), both summarizing past research and conducting original analysis. Collectively, the chapters constitute the first comprehensive evaluation of federal environmental justice policy as implemented over the past twenty years. As the title of the book suggests, the overall conclusion is that the federal government has failed to live up to most of the promises articulated in EO 12898 and in subsequent policy commitments. The news is not all bad, however. Recent efforts at the EPA suggest a reawakening of sorts with regard to federal attention to the disproportionate burdens confronting many poor and minority communities across the country. For the first time since the signing of EO 12898, the agency is taking significant steps to translate the intent of the executive order into specific policies and procedures in areas such as rulemaking, permitting, and enforcement. Although success is by no means assured, there is some reason for optimism, as many of the contributors in this volume make clear.

The rest of this introductory chapter has two objectives. First, to clearly set the stage for the analysis that follows, I briefly review the empirical literature that has developed over the past three decades to document the type and degree of race- and class-based disparities in environmental amenities. Second, I more fully articulate the approach taken in this book to evaluate the performance of the federal government, particularly the EPA, in integrating environmental justice considerations into its decision making.

Scholarly Inquiry into Environmental Justice

Scholarly interest in environmental justice dates back at least to the publication of *Toxic Wastes and Race in the United States* in 1987.[6] This study, completed by the United Church of Christ Commission on Racial Justice (1987), was the first national-scale study of the demographic correlates of hazardous waste location. The report examined the 415 commercial hazardous waste facilities operating at that time in the United States, as well as about 18,000 uncontrolled waste sites that the EPA had begun to catalog as part of the Superfund waste cleanup program. Among the main findings of the report was that race was the most significant variable of those examined in association with the location of commercial hazardous waste facilities. The analysis found that communities (defined as postal ZIP codes) with a commercial hazardous waste facility had on average about twice the percentage of minorities than communities without such a facility. The study reached qualitatively similar findings regarding the uncontrolled waste sites it analyzed.

The original *Toxic Wastes and Race in the United States* report was not methodologically sophisticated by modern-day social science standards, and it is important to note what the report did and did not establish. The findings were based primarily on statistical tests of differences in group means, and it did not include a complete multivariate analysis to control for potentially confounding factors. The study also focused on just two dimensions of hazardous waste. The report is most often cited for its findings regarding commercial treatment, storage, and handling facilities (or TSDFs, for short) that process waste from off-site sources. TSDFs are symbolically important because they are facilities that handle waste coming in from other facilities, often located in higher-income and predominantly white communities. It is this type of waste transfer that frequently leads to allegations of waste "dumping" in poor and minority communities. However, commercial TSDFs comprise just a small proportion of the facilities regulated under the RCRA, and most hazardous waste is treated on site at the same facility where it is generated. The uncontrolled waste sites considered in the analysis were diverse in the nature of the risks they posed to surrounding populations, and the study did not make any distinctions in this respect. More generally, the report neither demonstrated that residents of poor and minority neighborhoods faced elevated environmental risks, nor did it attempt to identify causal mechanisms.

These criticisms are by no means intended to diminish the importance of the report. What the study did establish was a statistically significant correlation between the location of active and inactive hazardous waste sites and contemporaneous ZIP code–level demographic attributes of the surrounding community, particularly its racial composition. And, regardless of any methodological shortcomings, *Toxic Wastes and Race in the*

United States was symbolically powerful and tremendously influential. The report provided additional motivation and impetus to the growing environmental justice movement. Community groups began to view their local struggles as part of a nationwide cause, and networks of environmental justice organizations began to form to share experiences and strategies for mobilization. In addition, a common and powerful narrative began to emerge that linked the location of hazardous waste sites to race, leading some civil rights leaders to charge "environmental racism" (Grossman 1994). In addition, the movement could point to the report as providing systematic and wide-scale evidence of race- and class-based disparities in environmental protection in their efforts to impress upon the federal government the need to take corrective action.

Of most importance for the purposes of this book, *Toxic Wastes and Race in the United States* also kicked off an enormous interdisciplinary project of scholarly inquiry. Although one cannot attribute all of the subsequent scholarship on environmental justice to this one report, there was not a broad research agenda on the subject before it. Research branched off in various directions. Scholars of philosophy and political theory picked up the subject to make normative arguments about the moral and ethical basis for rectifying environmental inequality. Legal scholars and social scientists (e.g., geographers, sociologists, public health experts, economists, and political scientists) began to investigate environmental justice as well, focusing mostly on documenting the presence and extent of inequities, and to a lesser extent on identifying the causes of and potential solutions to these inequities. The literature that examines patterns of environmental inequities is far too vast to summarize here.[7] Of the greatest relevance is a brief review of the empirical literature on the presence and causes of environmental inequities, as well as a recap of the little we know regarding the efficacy of the federal government's policy response.

The Search for Environmental Inequities

Empirical research on environmental justice has focused predominantly on the search for a better understanding of the degree and pattern of disparities in environmental outcomes. This research can generally be grouped into three categories of inequities: facility location, pollution levels, and policy implementation. The quality and sophistication of this research has improved immensely over the past twenty-five years. Scholars now have access to richer data on environmental hazards and community characteristics, and they are increasingly relying on tools such as geographic information systems (GIS), risk modeling, and careful statistical estimation to better understand the nature of disparities. Researchers are also giving more thought to the causal processes that explain these disparities. This is not to suggest that all questions have been answered; this is certainly not the case. Evidence remains mixed in some areas, but collectively, there is growing evidence of environmental inequities.

Facility location studies comprise the lion's share of the quantitative research on environmental justice.[8] Since the publication of *Toxic Wastes and Race in the United States*, hundreds of studies have examined whether various types of locally unwanted land uses are disproportionately located in minority and low-income communities. Many of these studies examine commercial hazardous waste TSDFs, the same types of facilities analyzed in the United Church of Christ report (e.g., Anderton et al. 1994; Been 1995; Been and Gupta 1997; Davidson and Anderton 2000; Mohai and Saha 2006, 2007; Oakes et al. 1996).[9] Other studies have expanded this basic line of inquiry to other types of potential environmental risks. Researchers, for example, have studied the characteristics of populations living near federal Superfund sites (e.g., Hamilton and Viscusi 1999; Hird 1993, 1994; Noonan 2008; Zimmerman 1993), and facilities that report toxic emissions as part of the federal Toxics Release Inventory (TRI) program (e.g., Cutter, Holm, and Clark 1996; Ringquist 1997; Pastor, Sadd, and Morello-Frosch 2004; Pollock and Vittas 1995). These studies vary widely in their research designs and methodological approaches (see Baden, Noonan, and Turaga 2007; Noonan 2008; Ringquist 2005), and by no means have they generated uniform results, with some studies finding no disparities whatsoever. But meta-analyses and other reviews of the literature have found good evidence to support claims of inequities in the locations of polluting facilities, particularly with regard to race (Brown 1995; Mohai and Bryant 1992; Ringquist 2005; Szasz and Meuser 1997).

The mere presence of a hazardous waste TSDF, a Superfund site, or a major source of air or water pollution does not necessarily mean that people living near these facilities face elevated public health risks. This fact is often overlooked in the broader environmental justice literature, especially in early studies of hazardous and solid waste landfills. To be sure, these types of unwanted land use often do create significant health risks for nearby populations, and even in cases where the risks are minimal, living in close proximity to industrial sites of any type often brings unwelcomed nuisances. These nuisances, such as noise and odor pollution and traffic congestion, reduce quality life regardless of their health risks. From a public policy perspective, however, it can be problematic to draw conclusions about the nature of environmental inequities from location studies alone, since the location of a facility mostly matters to the extent to which it leads to increased risk to nearby populations.

Recognizing this limitation in the literature, scholars have moved beyond location to examine whether racial and ethnic minorities and low-income populations also reside in areas with higher pollution levels. This line of research seeks to establish a more direct link between environmental risk and the population attributes of communities that host facilities. Several studies have found that minority and low-income communities are exposed to higher levels of air and water pollution than are communities with fewer minorities and higher income populations (e.g., Ash and Fetter 2004; Hird and Reese 1998; Pastor, Morello-Frosch, and Sadd 2006). Scholars have also examined the distribution of toxic emissions using TRI data. Here, too, there is evidence of at least modest race- and class-based inequity in the location of these releases (e.g., Arora and Cason 1999; Daniels and Friedman 1999; Downey 1998; Lester, Allen, and Hill 2001; Pastor, Morello-Frosch, and Sadd 2004; Perlin et al. 1995; Pollock and Vittas 1995), although some find evidence to the contrary (Holmes, Slade, and Cowart 2000). While most studies examine the correlation between pollution levels and community demographic characteristics, some scholars take the additional step of translating pollution levels into risk, many using the EPA's Risk-Screening Environmental Indicators model (Abel 2008; Abel and White 2011; Ash and Fetter 2004; Ash et al. 2012; Hipp and Lakon 2010; Pastor, Morello-Frosch, and Sadd 2004).

A third area where scholars have searched for environmental inequities is government implementation of pollution control laws. Advocates for environmental justice have often alleged that minority and low-income communities experience disproportionate environmental hazards in part as a result of differential treatment in policy implementation (Bryant 1995; Bullard 1993; Bullard and Johnson 2000; Collin 1993). Examples frequently pointed to include bias in permitting decisions, including neglecting to take into account cumulative risks for communities hosting multiple facilities, as well as unequal enforcement of major pollution control laws and slowness in remedial response to contaminated properties. Although this part of the literature is less well developed, several studies have sought to test these kinds of claims. Studies of cleanup decisions under the Superfund program, for instance, have found that EPA officials are less responsive to lower-income communities than wealthier communities, and they often adopt less stringent cleanup remedies (Hamilton and Viscusi 1999; Sigman 2001). Other studies have found that listing of sites to the National Priorities List (NPL), which makes them eligible for federal cleanup dollars, is less likely and generally slower when the sites are located in lower-income and minority communities (O'Neil 2007; Sigman 2001).

Other research on policy implementation has focused on regulatory enforcement of major environmental statutes, such as the CAA, the CWA, and the RCRA. These studies examine both EPA enforcement activities and those of state administrative agencies, given the large role that states play in enforcing federal pollution control laws. Some scholars have found that EPA and state regulatory officials conduct fewer inspections and impose fewer punitive sanctions when firms are located in poor and minority areas, but other studies have identified few such disparities (Earnhart 2004; Gray and Shadbegian 2004, 2012; Helland 1998; Konisky 2009a; Konisky and Schario 2010; Konisky and Reenock 2013b; Scholz and Wang 2006).

Collectively, this research on race- and class-based disparities in facility siting, pollution levels, and policy implementation focuses on identifying the presence of environmental inequities. In essence, this work seeks to uncover violations of environmental justice. With a few exceptions, however, this work does not seek to determine the causal mechanisms behind observed patterns of disparities. The search for causation is difficult, as many factors may play a role. Three explanations are commonly posited to explain observed environmental inequities, particularly in the context of the distribution of environmental risks.[10] First, of course, is intentional discrimination. Although often alleged by environmental justice advocates, such discrimination is extremely difficult to substantiate since it would require capturing the motivations of private- and public-sector decision makers. To date, there is no systematic empirical evidence showing this type of intentional discrimination (Ringquist 2004), but again, finding such evidence is not easy.

A second explanation is often referred to as *neighborhood transition* or *minority move-in,* which refers to household market dynamics (this explanation does not pertain to differential implementation of environmental laws). The basic idea is that companies decide where to locate a given facility based on economic criteria, such as the cost of land, access to raw materials, transportation networks, and labor supply. Over time, the property values around these facilities decline due, in part, to the environmental risks (perceived or real) that they create. People with the financial means to depart leave for other neighborhoods and are replaced with low-income and minority groups without such means. This sequence of events produces a contemporary pattern of inequity, but the original siting decision was not itself discriminatory. Testing the neighborhood transition argument is challenging since it requires collecting historical information to capture conditions at the time a siting decision is made, and the few studies completed to date have found uneven evidence to support this argument (Been and Gupta 1997; Oakes et al. 1996; Pastor, Sadd, and Hipp 2001; Saha and Mohai 2005; Wolverton 2009, 2012).

A third common explanation for inequitable environmental outcomes is political capacity, which is related to the tendency for poor and minority groups to have fewer resources to overcome collective action problems. In the present context, this means these groups often have less capacity to mobilize opposition to locally unwanted land uses or to pressure government into enforcing environmental laws more strictly. Corporate officials take this fact into account when deciding where to locate a facility or whether to comply with environmental obligations, as do government regulators when they set priorities or decide how to allocate their limited resources. Several studies have shown evidence to support the political capacity argument. John Hird and Michael Reese (1998) found political mobilization resources to be positively correlated with lower pollution levels, while James Hamilton (1995) found that firms were less likely to expand hazardous waste facilities in more politically active areas. In a study of the pace of site cleanup, Hilary Sigman (2001) found that sites located in politically active communities moved more quickly through the Superfund cleanup pipeline. Wayne Gray and Ron Shadbegian (2012) found that higher levels of regulatory activity were directed toward regulated facilities in politically active communities, which is consistent with my recent work with Chris Reenock (Konisky and Reenock 2013a, b), which found that regulatory officials pursued noncompliance in facilities less aggressively when they were located in some minority and lower-class communities. Further research to uncover patterns of disparities, as well as their origins, is certainly needed. Although we have learned a great deal about the presence, magnitude, and causes of the disproportionate risks faced by vulnerable communities, we know much less about the efficacy of government efforts to redress them.

Evaluations of Government Policy
The federal government responded to early evidence of the presence of environmental inequities with a series of policy initiatives. These measures were unveiled during the first half of the 1990s, but twenty years later, it is striking how little systematic analysis has been performed to evaluate their efficacy. To be sure, there has been considerable commentary about various aspects of federal environmental justice policy, such as EO 12898 and the EPA's 1998 guidance on permitting. In addition, there have been several critical reports of EPA performance from organizations such as the General Accounting Office (GAO), the National Academy of Public Administration, the National Environmental Justice Advisory Council (NEJAC), and the EPA's own inspector-general (these reports are discussed in chapter 2, "The Federal Response to Environmental Inequality").

But the academic literature has not given much consideration to whether federal policy successfully led to the full integration of justice considerations into environmental decision making. Although there are plenty of normative critiques, there have only been a handful of objective, empirically based studies. A few studies have analyzed the degree to which environmental justice has been considered as part of federal rulemaking. Gerber (2002) reviewed final rules issued between 1994 and 2002 and found that EO 12898 was infrequently utilized in a proactive manner, and that it rarely had a demonstrable effect on rules themselves. A couple of studies have examined whether and how environmental justice issues have been addressed as part of formal regulatory impact analyses (RIAs), which are economic analyses that are required by President Ronald Reagan's Executive Order 12291 (EO 12291) for all major federal rules (rules having costs of $100 million or more). In one recent assessment, Spencer Banzhaf found that the EPA generally has not seriously evaluated distributional effects in RIAs. Banzhaf (2011, pp. 5–6) concluded that the "EPA has tended to stop at perfunctory, pro forma assertions that it is not creating or exacerbating an environmental justice," rather than conducting a thorough assessment of distributional effects. Shalini Vajjhala, Amanda Van Epps, and Sarah Szambelan (2008) reached a similar conclusion in their content analysis of 129 RIAs across three federal agencies [the EPA, the Department of Energy (DOE), and the Department of Transportation (DOT)]. Overall, they found only modest evidence that environmental justice played more than a cursory role, and not much evidence that this changed after the issuing of EO 12898.

Other studies have investigated the effects of federal environmental justice policy in two other domains: enforcement and community-capacity building. In an article examining state regulatory enforcement of the CAA, CWA, and RCRA, I found only modest evidence that state administrative agencies empowered by the EPA to carry out enforcement performed more enforcement actions in poor and minority communities after the signing of EO 12898 (Konisky 2009b). In fact, in some instances, states actually performed *less* enforcement after the federal environmental justice policies went into effect. Studying a different area of federal policy, Shalini Vajjhala (2010) found that EPA grants designed to build community capacity to address local environmental and public health challenges were only in part being directed to the low-income and minority areas most burdened with toxic emissions. Vajjhala's examination of the EPA's Environmental Justice Small Grants Program raised questions about whether these resources were being targeted to the communities most in need.

Purpose and Plan for the Book

Previous policy evaluations have provided some analysis of specific components of federal environmental justice policy, but in general, the scholarly literature has not devoted much attention to carefully examining the efficacy of the policies put in place. This book does just that. Much of the focus will be on the performance of the EPA because it is the lead agency responsible for carrying out the provisions of EO 12898 and because it has the most capacity to do so given its portfolio of environmental regulatory responsibilities. The key question is whether the agency has effectively implemented EO 12898 and the other related policies it has put forward since. Of particular interest is the degree to which environmental justice considerations have been successfully integrated into the agency's programs and policies.

On the surface, there are many things one could point to that might suggest meaningful EPA progress on environmental justice, ranging from establishing a dedicated office to coordinate agency activities to initiating new data collection and dissemination tools to increasing community outreach efforts. The test of policy effectiveness here is more specific. The chapters in this book take a hard look at how well the EPA has performed in incorporating environmental justice considerations into its day-to-day decision making throughout the agency's core activities, such as permitting, standard setting, economic analysis, and compliance assurance and enforcement. In this way, the book engages with the details of EPA regulatory activities to carefully analyze where environmental justice goals have been achieved and where the agency has fallen short.

The book also contributes to ongoing debates in the political science literature about the ability of U.S. presidents to orchestrate policy change through administrative action. Presidents have a number of administrative tools available to them, ranging from resources within the executive office itself and strategic use of appointments to the exercise of discretion under existing statutes and the use of executive orders. In recent decades, presidents of both political parties have increasingly turned to administrative action in pursuit of environmental policy objectives. Klyza and Sousa (2008) argue that the use of executive power in this area, particularly executive orders, is largely a response to congressional gridlock, and that it has been done with mixed success. The far-reaching analysis of EO 12898 conducted by the contributors to this book provides additional empirical evidence regarding the efficacy of pursuing policy change solely through an administrative strategy, an issue I return to in chapter 2.

Conceptual Framework
The chapters that follow are loosely organized around three widely recognized dimensions or principles of environmental justice: distributive justice, procedural justice, and corrective justice.[11] *Distributive justice* is

the idea that all members of society have the right to equal treatment, and that outcomes should be distributed fairly. In the present context, distributive justice means the equitable distribution of environmental burdens and benefits. *Procedural justice* is the notion that people have the right to be treated equally in political decision making regarding the distribution of goods and services. While distributive justice focuses on outcomes, procedural justice emphasizes the fairness of the decision-making process itself. *Corrective justice* also involves fairness, but specifically in how instances of lawbreaking are resolved. In the environmental justice context, corrective justice generally means both fairness in how punishments are doled out for violations of environmental and public health laws, and compensation for any damages inflicted.[12]

The activities of the EPA and other federal agencies do not fit neatly into these categories of distributive, procedural, and corrective justice, since many can be classified in more than one category. Take permitting as an example. The question of whether a major air emissions source gets a CAA operating permit or a commercial hazardous waste TSDF receives a RCRA operating permit includes issues of both distributive and procedural justice. It involves distributive justice in the sense that the implications of the potential risks associated with one of these facilities may fall disproportionately on a particular community, and procedural justice in the sense that these communities may not have equal access to the permitting decision-making process. One can think similarly about the siting of a new facility, the setting of new pollution control standards, or the carrying out of regulatory enforcement.

While most of the literature has sought to identify situations where these principles of environmental justice have been violated, the purpose of this volume is different. Here, we are interested in determining whether the policies put in place to address such violations—past, present, and future—have been effective. Given that the nature of the problems and processes creating these violations are multifaceted (and in many cases, decades in the making), it would be unreasonable to expect that federal action over a twenty-year period would have solved them completely— that is, produce equal outcomes. Moreover, to be clear, EO 12898 and other environmental justice policy mandates require only the *consideration* of equity in government decision making. They do not require that the EPA make decisions on equity grounds. This is important to emphasize, and it is the standard by which the EPA and, where appropriate, other federal agencies will be held in the analysis that follows.

Chapter Overview

Any analysis of a subject as large as this one cannot possibly cover all its dimensions. This book does not evaluate all components of federal environmental justice policy;[13] rather, it focuses on the key regulatory activities at the heart of the environmental protection system. To situate the overall analysis, in chapter 2, I provide a brief historical overview of the political, social, and academic roots of current government environmental justice policy activities. The chapter describes the main initiatives comprising federal policy, including the creation of the Office of Environmental Justice within the EPA, the establishment of the NEJAC, and, most important, the signing of EO 12898 by President Clinton in 1994. This mostly descriptive discussion serves two purposes: to establish the context of the empirical chapters to follow, and to review in more detail what we know to date about their efficacy.

In chapter 3, "Federal Environmental Justice Policy in Permitting," Eileen Gauna examines federal environmental justice policy in the area of facility permitting. She provides a detailed analysis of how environmental justice issues have been adjudicated in several venues that are important to permitting under major environmental statutes. In so doing, Gauna addresses important questions of both distributive and procedural justice. Specifically, she analyzes the decisions of the federal Environmental Appeals Board (EAB), the EPA body responsible for adjudicating administrative appeals of permitting decisions under the major laws that the agency implements. She argues that environmental justice challenges to new permits brought to the EAB on the grounds of disparate impacts to low-income and minority communities have largely been unsuccessful, mostly due to the EAB granting a high degree of deference to the technical expertise of EPA permit writers. Thus, to this point, there is little evidence that the EPA has integrated environmental justice concerns directly into permitting decisions, although today, the EAB does require as a matter of practice that EPA permit writers perform an environmental justice analysis and invite broad participation in permitting decisions. Gauna also examines recent policy developments, such as *Plan EJ 2014* and EPA implementation of Title VI of the federal Civil Rights Act, which may suggest some optimism for the future consideration of environmental justice issues in federal and state permitting decisions.

Chapters 4 and 5 turn to rulemaking. In chapter 4, "Assessing the EPA's Experience with Equity in Standard Setting," Doug Noonan provides a rich analysis of how environmental justice issues arise in the context of standard setting. He begins by examining the set of policy instruments (command-and-control regulation, market-based mechanisms, and information-based approaches) that the EPA can pursue to achieve its goals under the statutes it implements, and how instrument choice can create different types of distributive and procedural justice concerns. Next, he explores a variety of equity concerns that have arisen in EPA rulemaking in practice under a variety of pollution control laws, and argues that the EPA has largely fallen short of incorporating equity considerations as part of standard setting. Noonan concludes the chapter with analysis of recent progress, but notes that much is left to be accomplished.

In chapter 5, "Evaluating Environmental Justice: Analytic Lessons from the Academic Literature and in Practice," Ron Shadbegian and Ann Wolverton analyze the difficult challenges that arise when considering the environmental justice effects of federal rules and regulations. Specifically, the authors examine five important issues in any analysis of distributional implications of a new environmental standard: (1) the geographic scope of the analysis, (2) the identification of potentially affected populations, (3) the selection of a comparison group, (4) how to spatially identify effects on population groups, and (5) how to measure exposure or risk in an analysis. For each issue, they consider how it has been addressed in the scholarly literature, as well as in practice by the EPA, by considering five recent proposed or final rulemakings completed under various pollution control statutes. Shadbegian and Wolverton conclude that even though there has been a substantial uptick in the number of rules that consider environmental justice issues in their accompanying economic analysis, there remain significant analytical issues—regarding both data and methodology—to resolve before this becomes a routinized practice. This analysis, paired with the evaluation done by Noonan in Chapter 4, speaks to key dimensions of distributive justice in the EPA rule-making process.

Dorothy Daley and Tony Reames examine public participatory processes in chapter 6, "Public Participation and Environmental Justice: Access to Federal Decision Making." They describe and analyze the way that federal agencies have involved the public—particularly low-income and minority individuals and the groups that represent them—in environmental decision making, which is an important part of government efforts to achieve procedural justice. Their analysis is comparative, and it extends beyond the EPA to other federal agencies, including the DOE and the DOT. Among Daley and Reames's key findings are that there is significant variation in how public participatory processes are used to address environmental justice concerns across the three agencies examined, and that, while opportunities for public involvement have increased since the signing of EO 12898, actual participation from low-income and minority communities has been uneven.

Chapters 7 and 8 of the book turn to questions of corrective justice and, more specifically, of whether environmental laws are implemented and adjudicated fairly in regulatory enforcement and in the courts. In chapter 7, "Evaluating Fairness in Environmental Regulatory Enforcement," Chris Reenock and I examine patterns of race and class bias in federal and state regulatory enforcement using an original statistical analysis. This quantitative approach differs from the types of analyses conducted in the rest of the book, but it does so to take advantage of the rich store of historical data on EPA and state enforcement activity. Specifically, we study whether federal and state enforcement efforts under the CAA—inspections and punitive actions—changed in terms of poor and minority communities in the years following the adoption of the federal environmental justice policy initiative in the mid-1990s. EO 12898 and the EPA's environmental justice strategy documents explicitly called for the targeting of regulated pollution sources in these communities as a way to redress disproportionate environmental burdens. Our analysis, however, yields little evidence that government enforcement effort changed much during the post-policy period.

Elizabeth Gross and Paul Stretesky examine the role of the judicial system in addressing environmental inequalities in chapter 8, "Environmental Justice in the Courts." Although federal environmental justice policy did not (and could not) mandate that the judicial system address equity in environmental decision making, an evaluation of the courts is an important part of this volume. From the early years of the federal government's attention to environmental justice, there were explicit calls to use the courts to pursue fairness and corrective action pertaining to disproportionate burdens. Moreover, many in the environmental justice community alleged that the courts were unfair in their disposition of cases involving environmental violations. The courts, therefore, are viewed by many as a tool for achieving corrective justice and a mechanism by which environmental justice groups could hold the EPA and other federal agencies accountable for discriminatory practices.

In chapter 8, Gross and Stretesky first examine fairness in environmental litigation outcomes through a careful review of empirical studies on the subject. They then turn to the important question of whether courts can be successfully used by communities to overcome unfairness or bias in facility siting decisions by providing a venue for individuals and advocacy organizations to challenge what they believe to be discriminatory practices. They examine this question in the context of Title VI of the Civil Rights Act of 1964, the Equal Protection Clause of the Fourteenth Amendment to the U.S. Constitution, and the National Environmental Policy Act. They conclude with a set of lessons learned that can be employed to help achieve fairer outcomes in the future.

In the final chapter of the book, "Federal Environmental Justice Policy: Lessons Learned," I review the main findings from the preceding chapters. Collectively, the chapters point to a disappointing overall conclusion: the federal government, especially the EPA, has not effectively integrated environmental justice into the decision making of its core regulatory programs and activities. The reasons for this outcome pertain both to challenges specific to the areas studied—that is, permitting, standard setting, economic analysis, public participation, regulatory enforcement, and use of the courts—and to several factors that cut across these areas. Specifically, I identify three factors that have impeded general progress: failure to develop clear policy guidance, inadequate coordination across EPA regions and states, and inconsistent agency leadership. The challenge that the EPA faced to fully implement EO 12898 was difficult, but these impediments to progress are self-inflicted and could be remedied by future efforts. Although the book offers a sobering assessment of federal environmental justice policy, and a discouraging one for the EPA, there is some basis for optimism that current policy efforts could remedy some of these issues. The task remains difficult, and it will require sustained commitment from the EPA and other government agencies. But so long as environmental inequities exist, the task certainly demands our attention.

Notes

1. The business community, of course, is not monolithic in its approach to environmental issues, although as argued here, most industries generally do not seek stronger environmental protections. For a detailed discussion of business and the environment, see Kraft and Kamieniecki (2007).

2. For a recent and comprehensive treatment of this subject, see Layzer (2012).

3. The book will not use the term *environmental racism* synonymously with *environmental justice* or *environmental equity*. Environmental racism is used by some environmental justice advocates and scholars to characterize violations of environmental justice, but it builds into the definition a conclusion that disproportionate burdens result from explicit racial discrimination. As discussed later in the chapter, there is still active debate about the causal mechanisms underlying observed environmental inequities, and such outcomes may occur for reasons other than deliberate discrimination. More generally, this topic is beyond the scope of the book.

4. Of course, environmental justice issues are not unique to the United States. In addition to similar circumstances in other nations, there has been considerable attention paid to "North-South" environmental justice issues (see, for example, Carmin and Agyeman (2011) and Pellow (2007)).

5. "Letter, Circa Earth Day 1990," reprinted in Rechtschaffen and Gauna (2002).

6. Robert Bullard's (1983) study of solid waste site location in Houston is a notable exception that predated the United Church of Christ report.

7. Christopher Foreman (1998) has noted that contributors to this literature tend to have one of two primary aims: political agenda setting or empirical enlightenment. Foreman argues that contributors with political goals seek mostly "to dramatize and legitimize environmental justice as a problem," while those seeking empirical enlightenment are more interested in better understanding environmental justice phenomena. These literatures are not always distinct, and some authors are clearer than others in stating their objectives.

8. In addition to these quantitative studies, there are many excellent case studies of facility siting in particular places. Examples include David Pellow's (2002) analysis of waste landfills and incinerators in Chicago, Steve Lerner's (2004) study of chemical plants in Diamond, Louisiana, and Julie Sze's (2006) study of industrial plants in New York City. These detailed histories of the people and politics involved in local decisions provide valuable contributions to our understanding of how (and when) the interests of elected officials, community organizations, and corporate actors collide.

9. The United Church of Christ has also sponsored two follow-up studies to its original report, each of which found similar results using more advanced analysis (Goldman and Fitton 1994; Bullard et al. 2007).

10. Two other explanations sometimes offered suggest that the siting of facilities (and, thus, the pollution they cause) result from technical (geological characteristics of a site) and economic (proximity to raw materials, transportation corridors,

labor supply, cheap land, etc.) criteria. In essence, these explanations suggest that low-income and minority residents of these areas simply live in the wrong place, at the wrong time.

11. Scholars have put forward various frameworks to disaggregate the types of equity considerations at stake in environmental justice. Bullard (1994b), for example, has divided environmental justice issues into three equity categories: procedural, geographic, and social. Manaster (1995) has identified three concepts—distributive justice, corrective justice, and procedural justice. To these three, Kuehn (2000) added social justice, while others have developed slightly different combinations (e.g., Kaswan 1997; Taylor 2000).

12. This discussion is based on the definitions of distributive, procedural, and corrective justice provided by Kuehn (2000).

13. Among the important subjects that the book does not cover in any detail include research performed or funded by the agency focused on overcoming data gaps in the areas of environmental health disparities, the development of data and screening tools to assist in decision making, and efforts to build community capacity to address environmental risks. In addition, the book does not consider if and how environmental justice goals have been carried out through specific programs, such as brownfields redevelopment, job and other technical training, or home lead abatement.

References

Abel, Troy D. 2008. Skewed Riskscapes and Environmental Injustice: A Case Study of Metropolitan St. Louis. *Environmental Management* 42 (2): 232–248.

Abel, Troy D., and Jonah White. 2011. Skewed Riskscapes and Gentrified Inequities: Environmental Exposure Disparities in Seattle, Washington. *American Journal of Public Health* 101 (S1): S246–S254.

Adler, Nancy E., and Katherine Newman. 2002. Socioeconomic Disparities in Health: Pathways and Policies. *Health Affairs* 21 (2): 60–76.

Anderton, Douglas L., Andy B. Anderson, John Michael Oakes, and Michael R. Fraser. 1994. Environmental Equity: The Demographics of Dumping. *Demography* 31 (2): 229–248.

Arora, Seema, and Timothy Cason. 1999. Do Community Characteristics Influence Environmental Outcomes? Evidence from the Toxics Release Inventory. *Southern Economic Journal* 65 (4): 691–716.

Ash, Michael, and T. Robert Fetter. 2004. Who Lives on the Wrong Side of the Environmental Tracks? Evidence from the EPA's Risk Screening Environmental Indicators Model. *Social Science Quarterly* 85 (2): 441–462.

Ash, Michael, James K. Boyce, Grace Chang, and Helen Scharber. 2012. Is Environmental Justice Good for White Folks? Industrial Air Toxics Exposure in Urban America. *Social Science Quarterly* 94 (3): 616–636.

Baden, Brett M., Douglas S. Noonan, and Rama Mohana R. Turaga. 2007. Scales of Justice: Is There a Geographical Bias in Environmental Equity Analysis? *Journal of Environmental Planning and Management* 50 (2): 163–185.

Banzhaf, H. Spencer. 2011. Regulatory Impact Analyses of Environmental Justice Effects. *Journal of Land Use & Environmental Law* 27 (1): 1–30.

Been, Vicki. 1995. Analyzing Evidence of Environmental Justice. *Journal of Land Use & Environmental Law* 11 (1): 1–36.

Been, Vicki, and Francis Gupta. 1997. Coming to the Nuisance or Going to the Barrios? A Longitudinal Analysis of Environmental Justice Claims. *Ecology Law Quarterly* 24:1–56.

Brown, Phil. 1995. Race, Class, and Environmental Health: A Review and Systematization of the Literature. *Environmental Research* 69 (1): 15–30.

Browner, Carol. 1993. Statement of Carol M. Browner, Administrator, U.S. Environmental Protection Agency, before the Government Operations Committee, United States House of Representatives, May 6, p. 3–4.

Brulle, Robert J., and David N. Pellow. 2006. Environmental Justice: Human Health and Environmental Inequalities. *Annual Review of Public Health* 27:103–124.

Bryant, Bunyan. 1995. *Environmental Justice: Issues, Policies, and Solutions.* Washington, DC: Island.

Bullard, Robert D. 1983. Solid Waste Sites and the Black Houston Community. *Sociological Inquiry* 53:273–288.

Bullard, Robert D. 1993. Anatomy of Environmental Racism and the Environmental Justice Movement. In *Confronting Environmental Racism: Voices from the Grassroots*, ed. Robert D. Bullard, pp. 15–39. Boston: South End Press.

Bullard, Robert. D. 1994a. *Dumping in Dixie: Race, Class, and Environmental Quality*, 2nd Ed. Boulder, CO: Westview Press.

Bullard, Robert D. 1994b. Overcoming Racism in Environmental Decisionmaking. *Environment* 36 (4): 10–20, 39–44.

Bullard, Robert S., and Glenn S. Johnson. 2000. Environmental Justice: Grassroots Activism and Its Impact on Public Policy Decision Making. *Journal of Social Issues* 56 (3): 555–578.

Bullard, Robert D., Paul Mohai, Robin Saha, and Beverly Wright. 2007. *Toxic Wastes and Race at Twenty, 1987–2007.* Cleveland, OH: United Church of Christ, Justice and Witness Ministries.

Carmin, JoAnn, and Julian Agyeman. 2011. *Environmental Inequalities Beyond Borders: Local Perspectives and Global Injustices.* Cambridge, MA: MIT Press.

Clark, Lara P., Dylan B. Millet, and Julian D. Marshall. 2014. National Patterns in Environmental Injustice and Inequality: Outdoor NO2 Air Pollution in the United States. *PLoS ONE* 9 (4): e94431. doi:.10.1371/journal.pone.0094431

Collin, Robert W. 1993. Environmental Equity and the Need for Government Intervention: Two Proposals. *Environment* 35:41–43.

Cutter, Susan, Danika Holm, and Lloyd Clark. 1996. The Role of Geographic Scale in Monitoring Environmental Justice. *Risk Analysis* 16 (4): 517–526.

Daniels, Glynis, and Samantha Friedman. 1999. Spatial Inequality and the Distribution of Industrial Toxic Releases: Evidence from the 1990 TRI. *Social Science Quarterly* 80 (2): 244–257.

Davidson, Pamela, and Douglas L. Anderton. 2000. Demographics of Dumping II: A National Environmental Equity Survey and the Distribution of Hazardous Materials Handlers. *Demography* 37 (4): 461–466.

Davies, J. Clarence, and Jan Mazurek. 1998. *Pollution Control in the United States: Evaluating the System*. Washington, DC: Resources for the Future Press.

Downey, Liam. 1998. Environmental Injustice: Is Race or Income a Better Predictor? *Social Science Quarterly* 79 (4): 766–778.

Earnhart, Dietrich. 2004. Regulatory Factors Shaping Environmental Performance at Publicly-Owned Treatment Plants. *Journal of Environmental Economics and Management* 48 (1): 655–681.

Environmental Protection Agency (EPA). 2003. *Framework for Cumulative Risk Assessment. EPA/600/P-02/001F*. Washington, DC: U.S. Environmental Protection Agency, Office of Research and Development.

Environmental Protection Agency (EPA), Office of Inspector General. 2004. *EPA Needs to Consistently Implement the Intent of the Executive Order on Environmental Justice*. Report No. 2004-P-00007. Washington, DC. March 1.

Environmental Protection Agency (EPA). 2011. *Plan EJ 2014*. Washington, DC. http://www.epa.gov/environmentaljustice/resources/policy/plan-ej-2014/plan-ej-2011-09.pdf.

Foreman, Jr. Christopher H. 1998. *The Promise and Peril of Environmental Justice*. Washington, DC: Brookings Institution Press.

Gee, C. Gilbert, and Devon C. Payne-Sturges. 2004. Environmental Health Disparities: A Framework Integrating Psychosocial and Environmental Concepts. *Environmental Health Perspectives* 1123:1645–1653.

General Accounting Office (GAO). 2005. *EPA Should Devote More Attention to Environmental Justice When Developing Clean Air Rules*. GAO-05-289. July.

Gerber, Brian J. 2002. Administering Environmental Justice: Examining the Impact of Executive Order 12898. *Policy and Management Review* 2 (1): 41–61.

Goldman, Benjamin, and Laura Fitton. 1994. *Toxic Waste, and Race Revisited*. Washington, DC: Center for Policy Alternatives.

Gray, Wayne B., and Ronald J. Shadbegian. 2004. "Optimal" Pollution Abatement: Whose Benefits Matter, and How Much? *Journal of Environmental Economics and Management* 47:510–534.

Gray, Wayne B, and Ronald J. Shadbegian. 2012. Spatial Patterns in Regulatory Enforcement: Local Tests of Environmental Justice. In *The Political Economy of Environmental Justice*, ed. H. Spencer H. Banzhaf. Palo Alto: CA: Stanford University Press.

Grossman, Karl. 1994. The People of Color Environmental Summit. In *Unequal Protection: Environmental Justice and Communities of Color*, ed. Robert D. Bullard. San Francisco: Sierra Club Books.

Hamilton, James T. 1993. Politics and Social Costs: Estimating the Impact of Collective Action on Hazardous Waste Facilities. *Rand Journal of Economics* 24 (1): 101–125.

Hamilton, James T. 1995. Testing for Environmental Racism: Prejudice, Profits, Political Power. *Journal of Policy Analysis and Management* 14 (1): 107–132.

Hamilton, James T., and Kip Viscusi. 1999. *Calculating Risks? The Spatial and Political Dimensions of Hazardous Waste Policy*. Cambridge, MA: MIT Press.

Helland, Eric. 1998. The Enforcement of Pollution Control Laws: Inspections, Violations, and Self-Reporting.. *Review of Economics and Statistics* 80 (1): 141–153.

Hipp, John R., and Cynthia M. Lakon. 2010. Social Disparities in Health: Disproportionate Toxicity Proximity in Minority Communities over a Decade. *Health & Place* 16 (4): 674–683.

Hird, John A. 1994. *Superfund: The Political Economy of Risk*. Baltimore: Johns Hopkins University Press.

Hird, John A. 1993. Environmental Policy and Equity: The Case of Superfund. *Journal of Policy Analysis and Management* 12 (2): 323–343.

Hird, John A., and Michael Reese. 1998. The Distribution of Environmental Quality: An Empirical Analysis. *Social Science Quarterly* 79 (4): 693–716.

Holmes, A., Barrett A. Slade, and Lary Cowart. 2000. Are Minority Neighborhoods Exposed to More Environmental Hazards? Allegations of Environmental Racism. *Real Estate Review* 30 (2): 50–57.

Institute of Medicine. 2002. *Unequal Treatment: Confronting Racial and Ethnic Disparities in Health Care*. Washington, DC: National Academies Press.

Jackson, Lisa P. 2010. Seven Priorities for EPA's Future. Memorandum to All EPA Employees. January 12. http://blog.epa.gov/administrator/2010/01/12/seven-priorities-for-epas-future.

Kaswan, Alice. 1997. Environmental Justice: Bridging the Gap Between Environmental Law and "Justice." *American University Law Review* 47 (2): 221–301.

Klyza, Christopher McGrory, and David Sousa. 2008. *American Environmental Policy, 1990–2006: Beyond Gridlock*. Cambridge, MA: MIT Press.

Konisky, David M. 2009a. Inequities in Enforcement? Environmental Justice and Government Performance. *Journal of Policy Analysis and Management* 28 (1): 102–121.

Konisky, David M. 2009b. The Limited Effects of Federal Environmental Justice Policy on State Enforcement. *Policy Studies Journal: The Journal of the Policy Studies Organization* 37 (3): 475–496.

Konisky, David M., and Christopher Reenock. 2013a. Case Selection in Public Management Research: Problems and Solutions. *Journal of Public Administration: Research and Theory* 23 (2): 361–393.

Konisky, David M., and Christopher Reenock. 2013b. Compliance Bias and Environmental (In)Justice. *Journal of Politics* 75 (2): 506–519.

Konisky, David M., and Tyler S. Schario. 2010. Examining Environmental Justice in Facility-Level Regulatory Enforcement. *Social Science Quarterly* 91 (3): 835–855.

Kraft, Michael E., and Sheldon Kamieniecki. 2007. *Business and Environmental Policy: Corporate Interests in the American Political System*. Cambridge, MA: MIT Press.

Kuehn, Robert R. 2000. A Taxonomy of Environmental Justice. *Environmental Law Reporter* 30:10681–10703.

Layzer, Judith A. 2012. *Open for Business: Conservatives' Opposition to Environmental Regulation*. Cambridge, MA: MIT Press.

Lerner, Steve. 2004. *Diamond: A Struggle for Environmental Justice in Louisiana's Chemical Corridor*. Cambridge, MA: MIT Press.

Lester, James P., David W. Allen, and Kelly M. Hill. 2001. *Environmental Injustice in the United States: Myths and Realities*. Boulder, CO: Westview Press.

Manaster, Kenneth A. 1995. *Environmental Protection and Justice: Readings and Commentary on Environmental Law and Practice*. Cincinnati, OH: Anderson Publishing Company.

Mohai, Paul, and Bunyan Bryant. 1992. Environmental Racism: Reviewing the Evidence. In *Race and the Incidence of Environmental Hazards*, ed. Bunyan Bryant and Paul Mohai, 163–176. Boulder, CO: Westview Press.

Mohai, Paul, David Pellow, and J. Timmons Roberts. 2009. Environmental Justice. *Annual Review of Environment and Resources* 34:405–430.

Mohai, Paul, and Robin Saha. 2006. Reassessing Racial and Socioeconomic Disparities in Environmental Justice Research. *Demography* 43 (2): 383–399.

Mohai, Paul, and Robin Saha. 2007. Racial Inequality in the Distribution of Hazardous Waste: A National-Level Reassessment. *Social Problems* 54 (3): 343–370.

Morello-Frosch, Rachel, and Edmond D. Shenassa. 2006. The Environmental "Riskscape" and Social Inequality: Implications for Explaining Maternal and Child Health Disparities. *Environmental Health Perspectives* 114 (8): 1150–1153.

National Academy of Public Administration. 2001. "Environmental Justice in EPA Permitting: Reducing Pollution in High-Risk Communities Is Integral to the Agency's Mission." December.

Noonan, Douglas S. 2008. Evidence of Environmental Justice: A Critical Perspective on the Practice of EJ Research and Lessons for Policy Design. *Social Science Quarterly* 89 (5): 1154–1174.

Oakes, John Michael, Douglas L. Anderton, and Andy B. Anderson. 1996. A Longitudinal Analysis of Environmental Equity in Communities with Hazardous Waste Facilities. *Social Science Research* 25 (2): 125–148.

O'Neil, Sandra George. 2007. Superfund: Evaluating the Impact of Executive Order 12898. *Environmental Health Perspectives* 115 (7): 1087–1093.

Pastor, Manuel, Jr., Rachel Morello-Frosch, and James L. Sadd. 2006. Breathless: Schools, Air Toxics, and Environmental Justice in California. *Policy Studies Journal* 34 (3): 337–362.

Pastor, Manuel, Jr., Jim Sadd, and John Hipp. 2001. Which Came First? Toxic Facilities, Minority Move-In, and Environmental Justice. *Journal of Urban Affairs* 23 (1): 1–21.

Pastor, Manuel Jr., James L. Sadd, and Rachel Morello-Frosch. 2004. Waiting to Inhale: The Demographics of Toxic Air Release Facilities in 21st-Century California. *Social Science Quarterly* 85 (2): 420–440.

Pellow, David N. 2002. *Garbage Wars: The Struggle for Environmental Justice in Chicago.* Cambridge, MA: MIT Press.

Pellow, David N. 2007. *Resisting Global Toxics: Transnational Movements for Environmental Justice.* Cambridge, MA: MIT Press.

Perlin, Susan A., R. Woodrow Setzer, John Creason, and Ken Sexton. 1995. Distribution of Industrial Air Emissions by Income and Race in the United States: An Approach Using the Toxic Release Inventory. *Environmental Science & Technology* 29 (1): 69–80.

Pollock, Philip III, and M. Elliot Vittas. 1995. Who Bears the Burdens of Environmental Pollution? Race, Ethnicity, and Environmental Equity in Florida. *Social Science Quarterly* 76 (2): 294–310.

Rechtschaffen, Clifford, and Eileen Gauna, eds. 2002. *Environmental Justice: Law, Policy, and Regulation.* Durham, NC: Carolina Academic Press.

Ringquist, Evan J. 2005. Assessing Evidence of Environmental Inequities: A Meta-Analysis. *Journal of Policy Analysis and Management* 24 (2): 223–247.

Ringquist, Evan J. 2004. Environmental Justice. In *Environmental Governance Reconsidered: Challenges and Opportunities*, eds. Robert F. Durant, Daniel J. Fiorino, and Rosemary O'Leary. Cambridge, MA: MIT Press.

Ringquist, Evan J. 1997. Equity and Distribution of Environmental Risk: The Case of TRI Facilities. *Social Science Quarterly* 78 (4): 811–829.

Saha, Robin, and Paul Mohai. 2005. Historical Context and Hazardous Waste Facility Siting: Understanding Temporal Patterns in Michigan. *Social Problems* 52 (4): 618–648.

Scholz, John T., and Cheng-Lung Wang. 2006. Cooptation or Transformation? Local Policy Networks and Federal Regulatory Enforcement. *American Journal of Political Science* 50 (1): 81–97.

Sexton, Ken. 2000. Socioeconomic and Racial Disparities in Environmental Health: Is Risk Assessment Part of the Problem or Part of the Solution? *Human and Ecological Risk Assessment* 6:561–574.

Sigman, Hilary. 2001. The Pace of Progress at Superfund Sites: Policy Goals and Interest Group Influence. *Journal of Law & Economics* 44:315–344.

Szasz, Andrew, and Michael Meuser. 1997. Environmental Inequalities: Literature Review and Proposals for New Directions in Research and Theory. *Current Sociology* 45 (3): 99–120.

Sze, Julie. 2006. *Noxious New York: The Racial Politics of Urban Health and Environmental Justice*. Cambridge, MA: MIT Press.

Taylor, Dorceta E. 2000. The Rise of the Environmental Justice Paradigm: Injustice Framing and the Social Construction of Environmental Discourses. *American Behavioral Scientist* 43 (4): 508–580.

United Church of Christ Commission on Racial Justice. 1987. *Toxic Wastes and Race in the United States: A National Report on the Racial and Socioeconomic Characteristics of Communities with Hazardous Waste Sites*. New York.

Vajjhala, Shalini P. 2010. Building Community Capacity? Mapping the Scope and Impacts of the EPA Environmental Justice Small Grants Program. *Research in Social Problems and Public Policy* 18:353–381.

Vajjhala, Shalini P., Amanda Van Epps, and Sarah Szambelan. 2008. Integrating EJ into Federal Policies and Programs: Examining the Role of Regulatory Impact Analyses and Environmental Impact Statements. RFF Discussion Paper 08⊠45. Washington, DC: Resources for the Future.

Wolverton, Ann. 2009. Effects of Socio-Economic and Input-Related Factors on Polluting Plants' Location Decisions. *Berkeley Electronic Journal of Economic Analysis and Policy, Advances* 9 (1): http://ideas.repec.org/p/nev/wpaper/wp200808.html.

Wolverton, Ann. 2012. The Role of Demographic and Cost-Related Factors in Determining Where Plants Locate: A Tale of Two Texas Cities. In *The Political Economy of Environmental Justice*, ed. H. Spencer H. Banzhaf. Palo Alto: CA: Stanford University Press.

Yen, I. H., and S. Leonard Syme. 1999. The Social Environment and Health: A Discussion of the Epidemiologic Literature. *Annual Review of Public Health* 20:287–308.

Zimmerman, Rae. 1993. Social Equity and Environmental Risk. *Risk Analysis* 13 (6): 649–666.

2

The Federal Government's Response to Environmental Inequality

David M. Konisky

On February 11, 1994, President Bill Clinton issued Executive Order 12898 (EO 12898), requiring each federal agency to "make achieving environmental justice part of its mission by identifying and addressing, as appropriate, disproportionately high and adverse human health or environmental effects of its programs, policies, and activities on minority populations and low-income populations" (EO 12898, section 1–101). In signing the executive order, President Clinton declared that "[a]ll Americans have a right to be protected from pollution—not just those who can afford to live in the cleanest, safest communities. Today, we direct federal agencies to make environmental justice a part of all they do" (Benac 1994).

EO 12898 was signed with considerable fanfare from the Clinton administration, but it met mixed reactions from the environmental justice community. Among some onlookers, there was a palpable sense of optimism. Representative John Conyers, a Michigan Democrat representing the Detroit area, commented that the executive order "represents another significant step in the long march towards full equality and human dignity for minorities and low-income Americans." Conyers continued, "For too long, communities of African-Americans have been the dumping ground for pollution and public agencies have, at best, sat by. This executive order signals a beginning to the end of that horrible legacy" (Benac 1994). Deeohn Ferris, a spokesperson for the Lawyers' Committee on Civil Rights, noted that EO 12898 "is a good beginning. It goes a long way to galvanize the federal bureaucracy to address the environmental problems in the communities of low-income people and people of color" (Lee 1994). And Robert Bullard, one of the leading voices in the environmental justice community, remarked shortly after EO 12898 was signed that it is "very important that grassroots activists have got their issue all the way to the President. This sends a signal that environmental justice has to become integrated with all our environmental policies, including siting, enforcement, and public policy making" (Begley 1994).

Others viewed EO 12898 with considerable skepticism. On the day that President Clinton signed the order, Administrator Carol Browner of the U.S. Environmental Protection Agency (EPA) attempted to give a speech to members of the environmental justice community at a national conference in Arlington, Virginia. As Browner began to speak, she was met with loud protests. Several attendees staged a walkout, yelling "Wake up, EPA," a comment that Browner herself acknowledged by applauding and saying: "There are times I am not proud to be the head of the agency because we have not done what we should do to protect the people of this country. Once again, you were going to sit there and listen to a government bureaucrat tell you what the problem is" (McFarling 1994). Browner then abandoned her prepared remarks and turned the microphone over to conference participants for comments and questions. Many of the speakers relayed stories about the poor quality of air and drinking water and the high incidence of cancers and other diseases in their communities, and highlighted frustrations with what, in their view, was years of inaction by the EPA and other government agencies (McFarling 1994).

In many respects, EO 12898 represented the culmination of a sustained effort by a dedicated group of civil rights leaders, community organizers, and scholar-activists to push the federal government to respond to what they believed was overwhelming and indisputable evidence that minority and low-income groups faced disproportionate environmental risks. This was a hard-fought victory. Questions of equity were at that time (and some would argue still today) largely absent from the agendas of most national environmental advocacy organizations. The mainstream environmental community, led by groups such as the Sierra Club and the Natural Resources Defense Council, focused instead on fighting the regulatory and legal battles emerging from the implementation of the large set of federal environmental statutes passed in the preceding two decades, as well as on their more traditional domains of protecting land and biological resources. There was also a widely held perception that these groups did not represent minorities. In 1990, several environmental justice groups sent a letter to the ten largest national environmental advocacy groups criticizing them for, not only ignoring issues such as the toxic burdens in low-income and minority communities, but failing to hire minority staffers and board members (MacLachlan 1992; Sandweiss 1998). Twenty years later, many in the environmental justice advocacy community continue to lament the lack of diversity in national environmental organizations (Fears 2013, 2014).

Concerns about disproportionate environmental impacts surfaced from local community groups, some more organized than others, rising up against locally unwanted land uses (LULUs), most notably solid and hazardous waste landfills and incinerators. Although there were many previous instances of communities of color and low-income communities protesting LULUs, the set of events that galvanized the activist community occurred in Warren County, North Carolina, ten years before President Clinton signed EO 12898. Scholars of the history of environmental justice often point to the events in Warren County as the moment when concerns about the environment were first framed in a civil rights context. Robert Bullard (2000) notes that the environmental activism that emerged during this period came out of the "growing hostility to facility siting decisions that were seen as unfair, inequitable, and discriminatory toward poor people and people of color" (p. 29).

This hostility was on full display in Warren County in 1982. Four years earlier, North Carolina governor James Hunt had selected the small town of Afton for the location of a polychlorinated biphenyl (PCB) disposal site that was needed to bury more than 30,000 gallons of PCB-contaminated soil that had been illegally dumped along 200 miles of rural state roads. The "midnight dumping" was done by a trucking company that was attempting to evade a new EPA restriction on the resale of toxic soil. Governor Hunt claimed that the town of Afton was selected on technical grounds, a claim later refuted by William Sanjour, chief of the EPA's hazardous waste office at the time (Washington 2005).

Members of the community also believed differently. Afton, North Carolina, was a small, poor town, where 84 percent of the population was black. Warren County had the highest percentage of black population in the state and a poverty rate of about twice the state average. For many in the community, the decision by Governor Hunt was clearly driven by the town's demographics, not the technical suitability of the disposal site. Luther Brown, the pastor of the largest black congregation in Warren County, expressed this sentiment: "We know why they picked us. It's because it's a poor county—poor politically, poor in health, poor in education, and because it's mostly black. Nobody thought people like us would make a fuss" (Russakoff 1982).

Shortly following Governor Hunt's decision, residents created an organization, Warren County Citizens Concerned About PCBs, to protest the disposal site. Local residents, supported by the county chapter of the National Association for the Advancement of Colored People (NAACP), filed a lawsuit in federal court asking for an injunction on the grounds that the EPA's and the State of North Carolina's decision to site the PCB landfill in Afton represented racial discrimination due to disregarding alternative sites (*New York Times* 1982).

On August 4, 1984, a federal district judge denied the injunction, clearing the path for the completion of the landfill. In an effort to stop the transport of the contaminated soil a month later, local residents demonstrated, some lying down in the streets in front of the incoming trucks. The local protestors were joined by the leadership of some of the preeminent national civil rights groups, including Leon White of the United Church of Christ's Commission for Racial Justice, Benjamin Chavis, Joseph Lowery, and Fred Taylor of the Southern Christian Leadership Conference (SCLC), and Walter Fauntroy, the District of Columbia's delegate to the U.S. Congress and a member of the Congressional Black Caucus. In the end, hundreds of individuals were arrested as part of the three-week demonstration, but the state completed the removal and relocation of the contaminated soil to the landfill.

The protests in Warren County were not qualitatively different than the Not In My Backyard (NIMBY) protests occurring in many other parts of the country during this period. Neighborhood associations from across the country were coming together to organize against local environmental harms, real and perceived. Inspired by the example set by Lois Gibbs and the residents of Niagara, New York, in the Love Canal episode, people from across the country had heightened sensitivities to the potential harms associated with all sorts of contaminated sites and newly proposed facilities. However, a key difference regarding the events in Warren County was that the community affected viewed the reason for the disposal decision as one of deliberate racial discrimination, not just corporate negligence or government mismanagement.

Although the protests were unsuccessful in their immediate goal of stopping the landfill, the events in Warren County received widespread attention. The PCB contamination and the subsequent demonstrations received repeated coverage from national newspapers, such as the *New York Times* and the *Washington Post,* and were reported on during national nightly television news broadcasts.

The demonstrations in Warren County also prompted some of the earliest investigations of the patterns of environmental burdens facing poor and minority communities. Walter Fauntroy came away from the protests in Warren County with the suspicion that its circumstances were not unique. In December 1982, he requested that the U.S. General Accounting Office (GAO) conduct a study of the community characteristics of all offsite hazardous waste landfills operating in EPA's Region IV, which includes the eight southeastern states of Alabama, Florida, Georgia, Kentucky, Mississippi, North Carolina, South Carolina, and Tennessee. The GAO identified four such landfills: Chemical Waste Management (Sumpter County, Alabama), Industrial Chemical Company (Chester County, South Carolina), SCA Services (Sumpter County, South Carolina), and the PCB landfill in Warren County. The principal finding of the 1983 report was that African Americans comprised a majority of the population in three of the four communities, and more than a quarter of the population in each community had average income levels below the federal poverty line (General Accounting Office 1993).

Four years later, the United Church of Christ Commission for Racial Justice (UCC-CRJ) (1987) issued the report, *Toxic Wastes and Race in the United States.* Charles Lee, later a director of the EPA Office of Environmental Justice, was the principal author of this report, which represented the first nationwide study of the demographic correlates of hazardous waste facility locations. As discussed in chapter 1, the report examined the 415 commercial ("off-site") hazardous waste facilities operating at that time in the United States, as well as thousands of uncontrolled waste sites that were under review by the EPA as part of the Superfund program. Among the main findings of the report was that race was the most significant variable of those examined in association with the location of commercial hazardous waste facilities. *Toxic Wastes and Race in the United States* concluded with a detailed set of recommendations directed at governments at all levels to begin addressing disparate risks, as well as a call to churches and community organizations to raise awareness and dedicate resources to fighting against these risks.

The GAO and UCC-CRJ reports provided additional motivation to the burgeoning environmental justice movement and strengthened the developing narrative linking disproportionate environmental burdens and race. The UCC-CRJ report, in particular, explicitly framed the siting of commercial hazardous waste disposal facilities in terms of race. In his press conference at the National Press Club in Washington discussing the release of *Toxic Wastes and Race in the United States*, Ben Chavis, executive director of the UCC-CRJ, characterized the situation as "an insidious form of institutionalized racism. It is, in effect, environmental racism. ... Given the disproportionate effect of these wastes on racial and ethnic communities, this has become not only an environmental issue, but a racial justice issue as well" (PR Newswire 1987).[1]

The 1987 UCC-CRJ report was followed a few years later by a 1990 conference organized by the University of Michigan's School of Natural Resources. The Conference on Race and the Incidence of Environmental Hazards brought together scholar-activists to share their latest research on race-based disparities in environmental burdens. This conference proved to be partially an information-gathering event and partially an organizational event to develop a strategy to influence policymakers. Among the outcomes of the conference was a book that compiled the research presented, *Race and the Incidence of Environmental Hazards*, edited by the conference organizers, Bunyan Bryant and Paul Mohai. Following the conference, a small group of conferees, later dubbed the "Michigan Coalition," drafted a memo to political officials requesting meetings with Michael Deland, chair of the White House Council on Environmental Quality, William K. Reilly, administrator of the EPA, and Louis Sullivan, secretary of the U.S. Department of Health and Human Services. The memo called for widespread action by government agencies to incorporate equity concerns into their research, risk communication, personnel, and policy activities. The Michigan Coalition sent copies of this memo to all state governors, various state legislators, and to members of the Congressional Black Caucus (Bryant and Mohai 1992). As is discussed more later in this chapter, this memo went a long way toward inducing government action.

A year later, Chavis organized the first National People of Color Environmental Leadership Summit, which was held in Washington, D.C.[2] This event was attended by over 1,000 participants and provided an opportunity for community organizers from across the United States and abroad to share their experiences and their strategies for bringing about change. The summit is most often remembered for its adoption of seventeen "Principles of Environmental Justice." A couple of years later, Robert Bullard summarized these principles into a five-point framework: (1) protect all persons from environmental degradation; (2) adopt a public health prevention of harm approach; (3) place the burden of proof on those who seek to pollute; (4) obviate the requirement to prove intent to discriminate; and (5) redress existing inequities by targeting action and resources (Bullard 1996).

The summit solidified the national importance of the growing movement, expanded its scope far beyond an anti-toxics campaign, and redefined the overall agenda as one of securing environmental justice (Foreman 1998; Lester, Allen, and Hill 2001). Participants clearly saw the issue in increasingly broad terms. Bryant and Mohai (1992), writing at about that time, commented:

It is abundantly clear that new political winds are beginning to blow across the country, altering the environmental movement, as people of color take up their struggle under the environmental banner as an alternative way of giving recognition to their unsung history of fighting for justice and clean, safe neighborhoods. Confronted with massive exposure to hazardous waste, threatened by freeways, urban decay, or by huge urban development—surrounded by concrete streets, buildings, parking lots, and playgrounds, cutting them off from wilderness areas—people of color have positioned their struggle for economic and social justice squarely in the front seat of the environmental movement. (pp. 5–6)

These events, in addition to adding fuel to the environmental justice movement, set a new course for the government's consideration of these issues. Up to this point, government had not taken serious action to address the calls for reform. This was about to change.

The Government Response

By the early 1990s, the environmental justice movement had succeeded in getting the issue onto the agenda of the federal government.[3] On June 4, 1992, Democratic Representative John Lewis of Georgia, the veteran civil rights activist, introduced a bill titled the Environmental Justice Act of 1992. This short piece of legislation, based largely on a set of provisions first suggested by Chavis in testimony given before the House

Subcommittee on Health and the Environment on February 25, 1992, had the objective "To establish a program to assure nondiscriminatory compliance with all environmental, health and safety laws and to assure equal protection of the public health" (Environmental Justice Act of 1992). The crux of this proposed legislation was that the EPA was to identify "environmental high impact areas (EHIAs)," which the bill defined as the 100 counties in the country with the highest total weight of toxic chemicals over the preceding five-year period. Once identified, facilities in the EHIAs were to be inspected to check their compliance with environmental, health, and safety standards; individuals and organizations in these communities were to be designated eligible for grants; the EPA was to conduct studies on the acute and chronic human health effects on residents living in EHIAs; and there was to be a moratorium on the siting or permitting of any new toxic chemical facility in any EHIA for which the agency had determined that current facilities emitted chemicals in quantities that caused significant adverse health effects. In essence, the legislation would have made it difficult for new sources of toxic chemicals to be sited in these communities.

The Environmental Justice Act of 1992 had 32 cosponsors, half of whom were members of the Congressional Black Caucus, but the bill died in committee. Senator Al Gore, who would shortly become Clinton's vice president, introduced a companion bill in the U.S. Senate, but it also was not moved in committee. John Lewis reintroduced the legislation in later sessions of Congress, and similar bills have since been regularly introduced in both the U.S. House and Senate, but none have received serious consideration. Several environmental justice–related amendments to existing environmental statutes, such as the Resource Conservation and Recovery Act (RCRA) and the Comprehensive Environmental Response, Compensation, and Liability Act (CERCLA), better known as Superfund, have also been introduced, but these too have fallen short of becoming law.

The environmental justice community contemporaneously sought action from the executive branch of the federal government, particularly the EPA. One of the immediate outcomes of the efforts of the Michigan Coalition was that the EPA began to formally recognize and consider environmental inequality. Even before Administrator Reilly met with members of the Michigan Coalition, he had publicly commented on the findings reported at their January 1990 conference.[4] Later that year, Reilly established the Environmental Equity Workgroup, an internal agency commission to examine equity questions. The workgroup reached six principal conclusions. First, they determined that there were clear differences between racial groups in terms of disease and mortality. Second, they found that racial minority and low-income populations faced higher exposure to air pollutants, hazardous waste facilities, contaminated fish, and agricultural pesticides. Third, the workgroup found a paucity of environmental and health data disaggregated by race and income, and little information collected on the health risks posed by multiple industrial facilities, cumulative and synergistic effects, or multiple and different exposure pathways. Fourth, they found that the EPA could enhance its communication with racial minority and low-income groups about environmental problems. Fifth, the workgroup concluded that the agency's ten regional offices were well positioned to address equity concerns given their proximity to affected communities. Finally, the workgroup reaffirmed that Native Americans are a unique racial group with a special relationship with the federal government, as well as being a population that confronts distinct environmental problems (EPA 1992).

Although the workgroup stated its conclusions cautiously, emphasizing uncertainties and data limitations where they existed, it did deliver specific recommendations to the EPA: (1) increase its prioritization of environmental equity issues; (2) establish and maintain information that provides an objective basis for assessment of risks by income and race; (3) incorporate considerations of environmental equity into the risk assessment process; (4) identify and target opportunities to reduce high concentrations of risk to specific population groups; (5) where appropriate, assess and consider the distribution of projected risk reduction in major rulemakings and agency initiatives; (6) selectively review and revise permit, grant, monitoring, and enforcement procedures to address high concentrations of risk in racial minority and low-income communities and encourage states to do the same; (7) improve communication with racial minority and low-income communities and increase efforts to involve them in environmental policymaking; and (8) establish mechanisms, including a center of staff support, to ensure that environmental equity concerns are incorporated in its long-term planning and operations (EPA 1992).

Over the next few years, several actions were taken by the George H. W. Bush administration (and subsequently the Clinton administration) in response to these recommendations. In November 1992, the EPA created the Office of Environmental Equity, later renamed the Office of Environmental Justice (OEJ), to coordinate the agency's environmental justice efforts.[5] Initially a stand-alone office within the agency, the OEJ was institutionally relocated in 1995 to the Office of Enforcement and Compliance Assurance, where it remains today.[6] The OEJ has always been a small office; it currently has a professional staff of just 15 people, and the agency spent about $7.7 million on environmental justice activities in fiscal year 2012 (less than 1 percent of the total EPA budget).[7] In addition to this office, personnel in the EPA's main programmatic offices (e.g., air, water, pesticides) have environmental justice responsibilities, and each of the ten EPA regional offices has people working as environmental justice coordinators.

In 1993, the EPA established the National Environmental Justice Advisory Council (NEJAC) to obtain advice and recommendations from expert stakeholders outside the agency. NEJAC, similar to other federal advisory committees, is comprised of members representing diverse interests and serves in an advisory capacity only (i.e., its recommendations are nonbinding on the agency). Since its creation, NEJAC has held dozens of meetings and issued numerous reports, ranging in topics from the integration of environmental justice into permitting to broadening public participation in environmental decision making to disaster preparedness in the wake of Hurricane Katrina.[8]

This initial flurry of government action brings us back to February 1994, and President Clinton's signing of EO 12898, which established the first environmental justice-specific mandates at the federal level. The executive order was the most important administrative action on environmental justice taken by President Clinton, and it remains the hallmark federal action to address equity concerns. The principal objective set forth in EO 12898 was to ensure that federal agencies account for any disproportionately high and adverse human health effects on low-income and minority populations that could result from their own actions, including policy setting, implementation, and all other related activities.

EO 12898 called for the creation of an interagency working group on environmental justice to be directed by the EPA administrator, with membership consisting of the departmental heads of all of the major federal agencies. This working group was asked to coordinate federal action on environmental justice issues in the areas of strategy and research, and to help synchronize activities across the federal bureaucracy. EO 12898 also required each federal agency to develop an environmental justice strategy that: "identifies and addresses disproportionately high and adverse human health or environmental effects of its programs, policies, and activities on minority populations and low-income populations" (Executive Order 12898, section 1). In this respect, EO 12898 called for the federal government to take widespread action to incorporate equity concerns into its environmental decision making. Although the EPA was designated as the lead agency—and has remained in this position since—the executive order called for each federal agency to consider equity issues. EPA Administrator Carol Browner made note of this at the time: "This is a problem that cannot be solved by any one agency. The executive order will be an important step to solving the problem, but it is going to take a lot of work with a lot of people" (Cushman 1994).

In addition to calling for agencies to develop their own environmental justice strategies, EO 12898 also emphasized the need for additional data collection and analysis. In particular, it called for gathering data on risks (including cumulative risks) to human health and the environment, disaggregated by race, national origin, and income (Executive Order 12898, section 3). The inclusion of this provision reflected recognition that there was a lack of the type of data needed to fully understand whether the disproportionate effects perceived by low-income and racial minorities were borne out by the facts. A final noteworthy component of EO 12898 was that it called for more openness and attentiveness to the environmental justice concerns of the public (Executive Order 12898, section 5). This provision was a direct response to calls for greater transparency and inclusiveness in government decision making, something which environmental justice advocates had long sought.

It is important to emphasize that the executive order itself did not commit resources to fulfilling its requirements. Moreover, EO 12898 neither mandated that decisions be made with equity as a specific criterion (just that it be considered) nor invoke any specific statutory authority. And the executive order included specific language (as do many executive orders) that emphasized the limits of judicial review.[9] For these reasons, some scholars characterize the executive order itself as little more than a symbolic policy statement (Cooper 2001).

That said, to flesh out the provisions and intent of EO 12898, President Clinton issued a companion memorandum to the heads of the affected federal departments and agencies, in which he stated that EO 12898

is designed to focus Federal attention on the environmental and human health conditions in minority communities and low-income communities with the goal of achieving environmental justice. That order is also intended to promote non-discrimination in Federal programs substantially affecting human health and the environment, and to provide minority communities and low-income communities access to public information on, and an opportunity for public participation in, matters relating to human health or the environment. (Clinton 1994)

A central purpose of this memorandum was to highlight ways in which existing statutes might be used to achieve some of the executive order's goals. In particular, Clinton pointed to Title VI of the Civil Rights Act of 1964 and the National Environmental Policy Act of 1970 (NEPA). Regarding Title VI, the memorandum noted that agencies of the federal government should be sure that "all programs or activities receiving Federal financial assistance that affect human health or the environment do not directly, or through contractual or other arrangements, use criteria, methods, or practices that discriminate on the basis of race, color, or national origin." Although the language here was vague (perhaps deliberately so), this was a potentially far-reaching statement, and one that would be the subject of considerable controversy later in the administration, as the Department of Justice (DOJ) tried to determine exactly how Title VI could be used in this context (discussed later in chapters 3 and 8).

The memorandum was a little more specific with respect to how environmental equity concerns were to be integrated into the NEPA process. The NEPA statute requires the consideration of the impacts of all major federal actions that significantly affect the quality of the environment. Formally, agencies must prepare an environmental impact statement that outlines alternatives prior to moving forward with such actions.[10] NEPA does not require an agency to select the alternative with the least environmental impact; it would just need to consider such alternatives. In this way, the statute sets forth procedural requirements but does not mandate particular outcomes. As part of the NEPA process, agencies were now expected to consider "the effects on minority communities and low-income communities" as well. In addition, the Clinton memorandum directed that federal agencies should provide greater opportunities for community input in NEPA decisions. Details in the memorandum were lacking in terms of exactly how implementation of EO 12898 would occur, and specific guidance on how Title VI and NEPA were not released until President Clinton's second term.

The EPA also put forward an agencywide environmental justice strategy, enumerating a diverse set of measures that the agency had already initiated and would take in the future to implement EO 12898 (EPA 1995). Taken at face value, the strategy document indicated a genuine intent to fully incorporate equity considerations into agency decision making, and the EPA issued an implementation plan a year later to articulate how it would carry out the commitments made in the strategy (EPA 1996). The strategy stated unequivocally that its purpose was to "ensure the integration of environmental justice into the Agency's programs, policies, and activities consistent with the Executive Order" (EPA 1995). This broad statement was consistent with the declaration from Carol Browner that environmental justice was to be one of her four top priorities during her tenure as administrator. Collectively, these actions at the EPA seemed to signify a serious effort by the agency to bring attention to equity issues, and a commitment to integrate them into its decision making.

It is worth emphasizing that the response from the federal government during these formative of years of policymaking was limited almost exclusively to the executive branch. As was discussed earlier in the chapter, several pieces of legislation were introduced in each chamber of Congress, but none were given serious consideration. This has remained true in the years since.[11] Moreover, there have been only a handful of congressional hearings explicitly addressing environmental justice issues, most of which occurred in the U.S. House of Representatives amid the initial flurry of federal policy interest in 1993, and *before* the signing of EO 12898.[12] The U.S. Senate did not hold its first hearing on environmental justice until 2007.[13] As a consequence, evaluations of federal environmental justice policy must focus on the administrative actions that followed from the executive order, although the lack of a legislative mandate has in itself been an impediment to progress, a subject I return to in chapter 9.

The Effectiveness of Government Action after EO 12898

EO 12898 was the centerpiece reform put in place by the Clinton administration to promote environmental justice. The executive order embodied many important elements of distributive, procedural, and corrective justice, and it set the tone for the federal agenda on the issue. Many provisions contained in EO 12898 had been recommended by the EPA's Environmental Equity Workgroup two years prior, and many had been previously pushed for by activists and scholars in the advocacy community. In this respect, EO 12898 was a major achievement, representing both an official recognition of the importance of environmental justice and a call to action to the federal government to engage in a serious and systematic way with the issue.

The executive order called for a number of actions within both the EPA and the rest of the federal bureaucracy.[14] From standard setting and regulatory impact analysis to permitting and enforcement, EO 12898 and the policy statements that followed directed the federal government to make equity a central consideration in its decision making. Did the federal government, and particularly the EPA, follow through on these directives? The remaining chapters of this book will analyze this question in the context of the important procedural and regulatory actions that the EPA—and, where appropriate, other federal and state agencies—have (or have not) taken in the years that have followed. The book focuses mostly on the EPA for a couple of reasons. First, the EPA is the lead agency in implementing the executive order, and as such, it has special obligations to be sure that it follows the intent of the order. The EPA administrator is the chair of the interagency working group assigned to manage federal environmental justice activities, which in essence gives the agency the lead voice on this subject within the vast federal bureaucracy.

Second, the EPA "holds the keys" to the regulatory regime that can best address the equity concerns at the heart of environmental justice concerns. The EPA is the agency responsible for implementing the programs that comprise the U.S. pollution control system, be it the Clean Air Act (CAA), the Clean Water Act (CWA), the RCRA, or the CERCLA. The agency, then, either through its own actions at EPA headquarters, through the actions taken in its ten regional offices, or through its oversight of state environmental agencies, is responsible for issuing permits, setting and analyzing new standards, enforcing laws, etc. Other federal agencies are relevant as well, as their activities have consequences for environmental quality, but it is the EPA that matters the most for the environmental equity issues at stake.[15]

What should we expect in terms of the impact of EO 12898 on environmental decision making within the federal government? In addressing this question, it is important to keep in mind that there is a sizeable amount of literature arguing that presidents can effectively pursue policy objectives through the "administrative presidency" (Nathan 1983; Durant 1992). Executive orders, in particular, can be powerful tools used by presidents to "make laws," as Yale law professor Donald Elliott describes it (Elliott 1989). Research in political science has shown that presidents issue executive orders to shape policy and that the frequency of their use is related to features of their external political environment, such as the legislative environment in Congress, popularity with the public, macroeconomic conditions, and management challenges related to the federal bureaucracy (Howell 2003; Krause and Cohen 1997; Mayer 1999, 2002; Moe and Howell 1999).

In the realm of environmental policy in particular, there is considerable evidence that presidents of both political parties have routinely used executive orders to pursue policy goals. In a historical analysis on the use of environment-focused executive orders, West and Sussman (1999) found that they were consistently used by presidents, from Franklin D. Roosevelt to Clinton, mostly to implement congressional statutes, but also to reorganize federal agencies and to establish new policy initiatives.[16] Klyza and Sousa (2008) find a similar pattern when they updated these data through the George W. Bush administration.

Many of these executive orders have had great effect, some by expanding environmental protections and others by contracting them. For example, under the Antiquities Act of 1906, most presidents (all but Nixon, Reagan, and George H. W. Bush) have signed executive orders setting aside land for protection as "national monuments." In recent years, presidential use of the Antiquities Act has come in response to the unwillingness of Congress to use its own authority under the law (Klyza and Sousa 2008). On the other end of the spectrum, shortly after assuming office, President Reagan issued Executive Order 12291 (EO 12291), which required the completion of a "regulatory impact analysis" (i.e., a cost-benefit analysis) for major new rules issued by regulatory agencies. This executive order was designed to reduce the burden of regulations on the U.S. economy, particularly those related to health, safety and the environment. Among other requirements, EO 12291 required that, to the extent permissible by law, "Regulatory action shall not be undertaken unless the potential benefits to society for the regulation outweigh the potential costs to society."[17] The use of these types of executive orders—whether to expand or cut back environmental protections—is part of a growing tendency of presidents to use administrative actions to pursue environmental policy (Durant 1992; Klyza and Sousa 2008; Soden 1999).

Despite the evidence suggesting that executive orders have been effective tools for pursuing environmental policy goals, there is also some reason to be skeptical that EO 12898 generated far-reaching changes in EPA policies and programs. Although executive orders are unilateral, they are not unchecked. William Howell (2003), for example, has argued that Congress and the courts have some ability to counteract executive orders. More generally, the long-term impact of executive orders depends on the willingness of subsequent administrations to go along. This can be a significant challenge for executive orders such as EO 12898, which have the primary intent of mandating changes to the decision-making procedures of the federal bureaucracy. Executive orders can be modified or completely reversed by future presidents, and they often are (Waterman 2004). They can also simply be ignored. Unlike with failing to act on legislation, agencies cannot be sued for failing to implement executive orders or to live up to the promises of policy statements and strategy documents, and there is no formal mechanism by which outside stakeholders can induce compliance. For these reasons, we cannot simply assume that the various elements of EO 12898 were implemented, either by federal agencies through the duration of the Clinton administration or by the George W. Bush and Obama administrations that followed.

Several past reviews of EPA actions have concluded that the implementation of EO 12898, in fact, has been at best slow and inconsistent and at worst deeply flawed. For example, a 2001 report conducted by a panel of experts convened by the National Academy of Public Administration (NAPA 2001) found that the EPA was not successfully incorporating environmental justice concerns into its permitting programs for major pollution control laws (such as the CAA, the CWA, and the RCRA), even though the agency has significant statutory and regulatory authority to do so.[18] More generally, the NAPA experts concluded that "environmental justice has not yet been integrated fully into the agency's core mission or its staff functions," and that despite the commitment of senior EPA leadership, "[e]xpectations for specific outcomes have not accompanied these commitments, nor has the agency adopted methods for measuring progress in achieving them or accountability to ensure that EPA managers and staff work toward implementing environmental justice policies" (NAPA 2001, p. 2).

A report released by the EPA's own inspector general in March 2004 was also highly critical of the sluggish progress made by the agency up to that point. Among the many findings were the following:

• "EPA has not fully implemented Executive Order 12898 nor consistently integrated environmental justice into its day-to-day operations. EPA has not identified minority and low-income, nor identified populations addressed in the Executive Order, and has neither defined nor developed criteria for determining disproportionately impacted."

• "Although the Agency has been actively involved in implementing Executive Order 12898 for 10 years, it has not developed a clear vision or a comprehensive strategic plan, and has not established values, goals, expectations, and performance measurements."

• "In the absence of environmental justice definitions, criteria, or standards from the Agency, many regional and program offices have taken steps, individually, to implement environmental justice policies. This has resulted in inconsistent approaches by the regional offices. Thus, the implementation of environmental justice actions is dependent not only on minority and income status but on the EPA region in which the person resides." (EPA, Office of the Inspector General 2004, p. i).

Although the report targeted its critique specifically at the EPA under the George W. Bush administration, it is important to note that many of the criticisms reflected a decade of inattentiveness dating back to the Clinton administration. From the failure to define an "environmental justice community" and to develop useful national policy and guidance, the report concluded that the EPA had largely neglected to follow through on many facets of the executive order.

The GAO criticized the performance of the EPA in a 2005 report, finding that the agency was not considering equity in its rulemaking. It analyzed the development of three major air rules promulgated under the CAA (Government Accounting Office 2005): a 2000 rule to reduce sulfur in gasoline and emissions from new vehicles, a 2001 rule to reduce sulfur in diesel fuel and to reduce emissions from new heavy-duty engines, and a 2004 rule to implement a new ozone standard. The central conclusion reached in the report was that the EPA generally devoted little attention to environmental justice in the development of the rules. Among the key findings, the GAO found that senior managers responsible for developing the rules had not been directed to consider environmental justice, the agency workgroups charged with developing the rules had not received guidance or training on how to address environmental justice concerns, and the agency had analyzed environmental justice in only one of three economic analyses conducted as part of the rulemaking (and, even in this case, it was done in a way that inconsistently portrayed the relevant information about distributional impacts). The overall takeaway from the GAO report was that the agency was not appropriately evaluating equity in its rulemaking, and this general conclusion has been corroborated by other similar studies (Banzhaf 2011; Harper et al. 2013; Vajjhala, Van Epps, and Szambelan 2008).

These and other reviews indicate that the agency was neither fulfilling the specifics nor even the intent of EO 12898. The EPA was not alone in its slow and uneven response to the executive order. A study conducted by the Environmental Justice Law Professors Consortium and published in the *Environmental Law Reporter* in 2001 reached similar conclusions about other federal agencies (Binder et al. 2001). The legal scholars conducted a survey of federal agencies to gauge their commitment to environmental justice across a number of different activities. They concluded that "[i]ntegrating EJ [environmental justice] into program design has been relatively rare, and comprehensive assessment and analysis exceedingly uncommon. Based upon the agency responses, there appears to be only a few instances in which agencies have incorporated EJ principles and protections into programmatic design" (Binder et al. 2001, p. 22). The scholars also noted that the actions that were taken by federal agencies were generally modest, and that the most common approach to implementing the executive order was to repackage and identify existing programs and to initiate a small number of discrete new projects.

The slow implementation of EO 12898 is epitomized by the reluctance of the EPA to clearly define what constitutes an "environmental justice community." As contributors to later chapters of this book will attest, this is not an easy issue, and yet it is critical for an agency assigned to devote resources to ensuring that such communities are protected from disproportionate environmental burdens. (Similarly, the EPA has not clearly defined or established criteria for determining "disproportionate impact," another key term in the executive order.) In more concrete terms, the EPA has never settled on a definition of what constitutes a "minority" or "low-income" community for the purposes of EO 12898. In its 2004 assessment of the implementation of the executive order, the EPA inspector general concluded that the agency's "ability to comply with the Order's requirements in a consistent manner is impeded if it does not first identify the intended recipients of the environmental justice actions. Not defining what a minority or low-income community is makes it difficult for EPA program staff to incorporate environmental justice into its day-to-day activities" (EPA, Office of the Inspector General 2004, p. 8).

There are many other areas in which the EPA has been slow to move forward with the type of details needed for meaningful integration of environmental justice into its day-to-day decision making. For example, the EPA did not issue final guidance on how to incorporate environmental justice into NEPA reviews until 1998; the first toolkit for environmental justice assessment was not released until 2004; reviews of agency programs and activities were only first systemically conducted fifteen years after EO 12898; and comprehensive federal guidance on topics such as permitting, regulatory analysis, and enforcement have only just been completed as part of the Obama administration's *Plan EJ 2014* initiative (discussed later in this chapter).

In another area of policy, the EPA has long been criticized for its problematic enforcement of Title VI of the Civil Rights Act of 1964. As noted previously, provisions of this law prohibit recipients of federal financial assistance—such as state agencies issuing environmental permits under laws—from discriminating on the basis of race, color, or national origin in their programs or activities. Environmental justice advocates have brought scores of these cases to the EPA on these grounds (encouraged to do so by the memorandum issued by President Clinton that accompanied EO 12898), but the agency has neither processed them expeditiously nor, in the eyes of many critics, taken them seriously (Deloitte Consulting 2011; Mank 2008; Ringquist 2004; U.S. Commission on Civil Rights 2003). In addition, regulations issued by the EPA over the years concerning how the agency would implement Title VI have run into steep legal and political obstacles.

This initial look at the federal government's response to EO 12898 does not present a rosy picture, but one cannot draw firm conclusions based solely on what has been discussed to this point. Moreover, it is important to highlight some areas of clear success. The agency, for example, has implemented programs to build community capacity, particularly through the Environmental Justice Small Grants Program. Although some (e.g., Vajjhala Van Epps, and Szambelan 2008) have raised questions about whether the grants have been given to the communities that need them most, scores of local organizations have benefited from the program. The agency has also devoted significant efforts toward creating data and analytic tools to help the public make better use of information on pollution risks, facility performance, and EPA decision making. These are important accomplishments that should not go unrecognized. In general, however, federal attention to environmental justice declined during the second term of the Clinton administration and during the eight years of the George W. Bush administration. Although it would be an exaggeration to say that issues of environmental inequality completely fell off the agenda of the EPA and other federal agencies, the sense of urgency that accompanied the signing of EO 12898 and the subsequent policy statements clearly waned in the succeeding years.

A Policy Restart in the Obama Administration

This might be the end of the story if the clock stopped in 2008. Over the past several years, however, there has been a virtual restart of federal policy attention to environmental justice. When EPA Administrator Lisa Jackson took office in 2009, she emphasized that the agency "must take special pains to connect with those who have been historically underrepresented in EPA decision making, including the disenfranchised in our cities and rural areas, communities of color, native Americans, people disproportionately impacted by pollution, and small businesses, cities and towns working to meet their environmental responsibilities" (Jackson 2009).

A year later, Jackson named environmental justice one of seven priorities that the agency would pursue under her leadership. In doing so, she elevated environmental justice back to the top of the EPA's agenda, alongside problems such as climate change, air and water quality, and chemical safety (Jackson 2010). To fulfill this priority, the agency embarked on *Plan EJ 2014*, which the agency has described as a "roadmap to help EPA integrate environmental justice into its programs, policies, and activities" (EPA 2011, p. 4) Given that this was the stated objective of EO 12898 and prior EPA environmental justice strategies, it is worth noting that the need for *Plan EJ 2014* reflects a recognition by the agency itself that it had not successfully implemented the reforms of the mid-1990s.

Plan EJ 2014, timed to mark the twentieth anniversary of EO 12898, includes establishing new policy guidance, creating new assessment and information tools, and highlighting ways to build equity considerations into existing programs. The initiative has three major components: (1) cross-agency focus areas, relating to rulemaking, permitting, compliance and enforcement, community-based action and capacity building, and administrationwide programs (i.e., across the federal bureaucracy); (2) tool development areas, relating to science, law, information, and resources: and (3) program initiatives to illustrate models for making environmental justice part of the day-to-day operations of agency environmental protection efforts.

Plan EJ 2014 is a multiyear policy effort that the EPA believes has already yielded significant progress in some areas. In its February 2013 progress report, the agency highlighted the development of "EJ Legal Tools," a report that offers legal guidance on using existing environmental and civil rights statutes to pursue environmental justice. In addition, the EPA pointed to the development of a new screening tool, EJSCREEN, to help agency managers and staff identify geographic areas of concern, as well as the creation of new guidelines on incorporating environmental justice in decision-making processes, ranging from permitting to compliance and enforcement (EPA 2013b).

In some respects, *Plan EJ 2014* is similar to previous EPA environmental justice efforts, in that it reflects a purely administrative policy initiative. As such, there is no guarantee that it will be sustained by future presidential administrations (Democratic or Republican). But *Plan EJ 2014* is also distinct from prior EPA efforts in terms of its sheer comprehensiveness, and the agency has devoted significant resources to systematically outlining a strategy for fully integrating environmental justice into its policies and programs for the first time. In this sense, the agency is moving closer to fully implementing, albeit 20 years later, EO 12898. Given the infancy of *Plan EJ 2014*, the chapters in this volume cannot (and do not) evaluate the effectiveness of this new initiative. However, where appropriate, contributors to the book will discuss components of the initiative in the context of their findings, focusing on whether they will address the shortcomings of past policy efforts.

The intent of this book, instead, is to comprehensively examine how the federal government, particularly the EPA, has performed in carrying out both the spirit and the letter of EO 12898 and the other policy initiatives of the mid-1990s. The time is ripe for such an assessment, and the findings of the chapters can inform *Plan EJ 2014* and future policy efforts in this important area of environmental policy. The chapters that follow systematically evaluate the core regulatory activities of the EPA (and, when relevant, other federal agencies) to answer the question of whether (and how) environmental justice considerations have been integrated into decision-making outcomes and processes. The next chapter begins this analysis by looking at the area of permitting.

Notes

1. A few years later, Chavis provided a fuller definition of environmental racism: "racial discrimination in environmental policy-making, the enforcement of regulations and laws, the deliberate targeting of people-of-color communities for toxic waste facilities, the official sanctioning of the life-threatening presence of poisons and pollutants in our community and the history of excluding people of color from the leadership of the environmental movement" (Feinsilber 1991).

2. A second National People of Color Environmental Leadership Summit convened in October 2002 in Washington, DC, to commemorate ten years of accomplishments and to discuss directions for future activities.

3. The focus of this book is on the federal policy responses, but it is important to note that many state and local governments also took action to address environmental justice. Ringquist and Clark (1999, 2002) analyzed states' approaches to environmental equity issues and found that the degree of state efforts varied considerably, ranging from taking no action to commissioning statewide studies to establishing environmental justice advisory commissions to promulgating regulation and enacting legislation. Other studies have identified similar variation in state-led initiatives (Bonorris 2010; Lester, Allen, & Hill 2001; Targ 2005).

4. Speaking at Howard University on April 9, 1990, Reilly commented that the conference participants' review of the evidence "pointed out significantly disproportionate health impacts on minorities due to higher rates of exposure to pollution," quoted in Bryant and Mohai (1992, p. 5).

5. In addition to the creation of a stand-alone office, the EPA has also undertaken additional administrative reorganization. In 2002, the agency established a senior policy and leadership body, the Environmental Justice Steering Committee (now known as the Environmental Justice Executive Management Council), to facilitate and coordinate EPA action.

6. Some members of the environmental justice community were skeptical about the new office, fearing that its establishment outside of the EPA's main media (air, water, and waste) and enforcement units would institutionally marginalize equity issues. See Foreman (1998) for further discussion.

7. This is based on combining "Environmental Justice" line items listed under Enforcement and Superfund of the EPA's Fiscal Year 2014 budget summary (EPA 2013a).

8. Reports from the NEJAC can be found at http://www.epa.gov/compliance/environmentaljustice/nejac/recommendations.html.

9. Specifically, EO 12898 included the following provisions: "This order is intended only to improve the internal management of the executive branch and is not intended to, nor does it create any right, benefit, or trust responsibility, substantive or procedural, enforceable at law or equity by a party against the United States, its agencies, its officers, or any person. This order shall not be construed to create any right to judicial review involving the compliance or noncompliance of the United States, its agencies, its officers, or any other person with this order." ("Federal Actions to Address Environmental Justice in Minority Populations and Low-Income Populations," Executive Order 12898, section 6-609, February 11, 1994).

10. Environmental impact statements are required unless the agency determines that the forthcoming action would have no environmental impact based on predefined criteria (this is a categorical exclusion) or if the agency determines that there will be no significant impact through a formal evaluation. This is known as an *Environmental Assessment (EA)*.

11. A search of historical legislation at the U.S. Congress web site (congress.gov) using the keyword "environmental justice" turned up 124 bills that have been introduced in Congress since June 1992 (the date of the initial bills introduced by Representative Lewis and Senator Gore). Although some of these bills were only tangentially connected to environmental justice, many others called for meaningful action by the EPA and other federal agencies to achieve diverse environmental justice goals. Most of these bills were not acted upon after being introduced.

12. "Environmental Justice," Hearing before the Subcommittee on Civil and Constitutional Rights, Committee on the Judiciary, U.S. House of Representatives, March 3–4, 1993; "Environmental Issues," "Hearing before the Subcommittee on Transportation and Hazardous Materials, Committee on Energy and Commerce, U.S. House of Representatives, November 17–18, 1993; "Environmental Protec-

tion Agency Cabinet Elevation; Environmental Equity Issues," Hearing before the Subcommittee on Legislation and National Security, Committee on Government Operations, U.S. House of Representatives, April 28, 1993; "EPA's Title VI Interim Guidance and Alternative State Approaches," Hearing before the Subcommittee on Oversight and Investigations, Committee on Commerce, U.S. House of Representatives, August 6, 1998; "Environmental Justice and the Toxics Release Inventory Reporting Program: Communities Have a Right to Know," Hearing before the Subcommittee on Environment and Hazardous Materials, Committee on Energy and Commerce, U.S. House of Representatives, October 4, 2007.

13. "Oversight of EPA's Environmental Justice Programs," Hearing before the Subcommittee on Superfund and Environmental Health, Committee on Environment and Public Works, U.S. Senate, July 25, 2007.

14. EO 12898 specifically mentioned 11 federal agencies: the Department of Defense, the Department of Health and Human Services, the Department of Housing and Urban Development, the Department of Labor, the Department of Agriculture, the Department of Transportation (DOT), the DOJ, the Department of the Interior, the Department of Commerce, the Department of Energy (DOE), and the EPA. These agencies were also to be represented on the Interagency Working Group on Environmental Justice, along with officials from the Office of Management and Budget (OMB), the Office of Science and Technology Policy, the Office of the Deputy Assistant to the President for Environmental Policy, the Office of the Assistant to the President for Domestic Policy, the National Economic Council, and the Council of Economic Advisers.

15. One important exception is the implementation of the NEPA, which is the responsibility of the White House Council on Environmental Quality.

16. See Shanley (1983) for a detailed discussion of the use of executive orders on various environmental policy issues during the Nixon, Carter, and early Reagan administrations.

17. "Federal Regulation," Executive Order 12291, February 17, 1981. White House.

18. Lazarus and Tai (1999) reach similar conclusions about the EPA's authority to consider equity in permitting.

References

Banzhaf, H. Spencer. 2011. Regulatory Impact Analyses of Environmental Justice Effects. *Journal of Land Use & Environmental Law* 27 (1): 1–30.

Begley, Ronald. 1994. Environmental Justice Advanced by Clinton Executive Order. *Chemical Week*, p. 15, February 23.

Benac, Nancy. 1994. Clinton Signs "Environmental Justice" Order. *Associated Press*, February 11.

Binder, Denis et al. 2001. A Survey of Federal Agency Response to President Clinton's Executive Order No. 12898 on Environmental Justice. *Environmental Law Reporter* 31:11133.

Bonorris, Steven. 2010. *Environmental Justice for All: A Fifty State Survey of Legislation, Policies, and Cases.* 4th ed., San Francisco: American Bar Association and Hastings College of the Law.

Bryant, Bunyan, and Paul Mohai. 1992. *Race and the Incidence of Environmental Hazards: A Time for Discourse.*, 4–5. Boulder, CO: Westview Press.

Bullard, Robert D. 1996. *Unequal Protection: Environmental Justice and Communities of Color.* San Francisco: Sierra Club Books.

Bullard, Robert D. 2000. *Dumping in Dixie: Race, Class, and Environmental Quality.* Boulder, CO: Westview Press.

Clinton, William Jefferson. 1994. Memorandum on Environmental Justice. February 11.

Cooper, Phillip J. 2001. The Law: Presidential Memoranda and Executive Orders: Of Patchwork, Quilts, Trump Cards, and Shell Games. *Presidential Studies Quarterly* 31 (1): 126–141.

Cushman, John H., Jr. 1994. Clinton to Order Pollution Policy Cleared of Bias. *New York Times*, February 10.

Deloitte Consulting. 2011. Evaluation of the EPA Office of Civil Rights. Washington, DC: Deloitte Consulting. http://www.epa.gov/epahome/pdf/epa-ocr_20110321_finalreport.pdf.

Durant, Robert F. 1992. *The Administrative Presidency Revisited: Public Lands, the BLM, and the Reagan Revolution.* Albany, NY: State University of New York Press.

Elliott, E. Donald. 1989. Why Our Separation of Powers Jurisprudence Is So Abysmal. *George Washington Law Review* 57 (3): 506–532.

Environmental Justice Act of 1992, House Resolution 2105. Introduced May 12, 1993.

Environmental Protection Agency (EPA). 1992. *Environmental Equity: Reducing Risk For All Communities.* Volume 1. Workgroup Report to The Administrator. EPA230-R-92-008, pp. 25–31.

Environmental Protection Agency (EPA). 1995. *The EPA's Environmental Justice Strategy.*

Environmental Protection Agency (EPA). 1996. *1996 Environmental Justice Implementation Plan.* Washington, DC. EPA/300-R-96-004. April.

Environmental Protection Agency (EPA), Office of Inspector General. 2004. *EPA Needs to Consistently Implement the Intent of the Executive Order on Environmental Justice.* Report No. 2004-P-00007. Washington, DC. March 1.

Environmental Protection Agency (EPA). 2011. *Plan EJ 2014.* September. Washington, DC.

Environmental Protection Agency (EPA). 2013a. "FY 2014 EPA Budget in Brief." April. Washington, DC.

Environmental Protection Agency (EPA). 2013b. "Plan EJ 2014 Progress Report." February. Washington, DC.

Executive Order 12898 of February 11. 1994. Federal Actions to Address Environmental Justice in Minority Populations and Low-Income Populations. *Federal Register 59* (32).

Fears, Darryl. 2013. Within Mainstream Environmentalists Groups, Diversity Is Lacking. *Washington Post*, March 25.

Fears, Darryl. 2014. Study: History of Cultural Bias Has Led to a Lack of Diversity at Liberal Green Groups. *Washington Post*, August 10.

Feinsilber, Mike. 1991. Minorities Say Planners Ignore Them, But Affect Them Most. *Associated Press*. October 24.

Foreman, Jr. Christopher H. 1998. *The Promise and Perils of Environmental Justice*. Washington, DC: Brookings Institution Press.

General Accounting Office. 1983. *Siting of Hazardous Waste Landfills and Their Correlation with Racial and Economic Status of Surrounding Communities.* GAO/RCED-83-168.

Government Accountability Office. 2005. *EPA Should Devote More Attention to Environmental Justice When Developing Clean Air Rules. GAO-05-289.* July.

Harper, Sam et al. 2013. Using Inequality Measures to Incorporate Environmental Justice into Regulatory Analyses. *International Journal of Environmental Research and Public Health* 10 (9): 4039–4059.

Howell, William G. 2003. *Power Without Persuasion: The Politics of Direct Presidential Action*. Princeton, NJ: Princeton University Press.

Jackson, Lisa P. 2009. Opening Memorandum to EPA Employees:. Memorandum to All EPA Employees. January 23. http://blog.epa.gov/administrator/2009/01/26/opening-memo-to-epa-employees.

Jackson, Lisa P. 2010. Seven Priorities for EPA's Future. Memorandum to All EPA Employees. January 12, 2010. http://blog.epa.gov/administrator/2010/01/12/seven-priorities-for-epas-future.

Krause, George A., and David B. Cohen. 1997. Presidential Use of Executive Orders, 1953–1994. *American Politics Research* 25 (4): 458–481.

Klyza, Christopher McGrory, and David Sousa. 2008. *American Environmental Policy, 1990–2006: Beyond Gridlock*. Cambridge, MA: MIT Press.

Lazarus, Richard J., and Stephanie Tai. 1999. Integrating Environmental Justice into EPA Permitting Authority. *Ecology Law Quarterly* 26:617–678.

Lee, Gary. 1994. Clinton Executive Order Gives Boost to Mission. *Washington Post*, February 17.

Lester, James P., David W. Allen, and Kelly M. Hill. 2001. *Environmental Injustice in the United States*. Boulder, CO: Westview Press.

MacLachlan, Claudia. 1992. Tension Underlies Rapport With Grassroots Groups. *National Law Journal*, September 21, S10.

Mank, Bradford C. 2008. Title VI. In *The Law of Environmental Justice: Theories and Procedures to Address Disproportionate Risks*, ed. Sheila R. Foster, 23–65. Chicago: American Bar Association.

Mayer, Kenneth R. 2002. *With the Stroke of a Pen: Executive Orders and Presidential Power*. Princeton, NJ: Princeton University Press.

Mayer, Kenneth R. 1999. Executive Orders and Presidential Power. *Journal of Politics* 61 (2): 445–466.

McFarling, Usha Lee. 1994. Poor, Minorities Seek Role in "Environmental Justice"; Community Activists Take Over EPA-Sponsored Session. *Boston Globe,* February 13.

Moe, Terry M., and William G. Howell. 1999. The Presidential Power of Unilateral Action. *Journal of Law Economics and Organization* 15 (1): 132–179.

National Academy of Public Administration. 2001. Environmental Justice in EPA Permitting: Reducing Pollution in High-Risk Communities is Integral to the Agency's Mission. December.

Nathan, Richard P. 1983. *The Administrative Presidency*. New York: Wiley.

New York Times. 1982. Carolinians Angry Over PCB Landfill, August 11.

PR Newswire. 1987. Civil Rights Head Charges 'Environmental Racism.' April 16.

Ringquist, Evan J. 2004. Environmental Justice. In *Environmental Governance Reconsidered: Challenges and Opportunities*, eds. Robert F. Durant, Daniel J., Fiorino, and Rosemary O'Leary, 235–287. Cambridge, MA: MIT Press.

Ringquist, Evan J., and David H. Clark. 1999. Local Risks, States' Rights, and Federal Mandates: Remedying Environmental Inequities in the U.S. Federal System. *Publius: The Journal of Federalism* 29 (2): 73–93.

Ringquist, Evan J., and David H. Clark. 2002. Issue Definition and the Politics of State Environmental Justice Policy Adoption. *International Journal of Public Administration* 25 (2&3): 351–389.

Russakoff, Dale. 1982. As in the '60s, Protestors Rally; But This Time the Foe is PCB. *Washington Post,* October 11.

Sandweiss, Stephen. 1998. The Social Construction of Environmental Justice. In *Environmental Injustices, Political Struggles*, ed. David E. Camacho, 38–58. Durham, North Carolina: Duke University Press.

Shanley, Robert A. 1983. Presidential Executive Orders and Environmental Policy. *Presidential Studies Quarterly* 13 (3): 405–416.

Soden, Dennis L. 1999. *The Environmental Presidency*. Albany: State University of New York Press.

Targ, Nicholas. 2005. The States' Comprehensive Approach to Environmental Justice. In *Power, Justice, and the Environment*, ed. David Naguib Pellow and Robert J. Brulle, 171–184. Cambridge, MA: MIT Press.

United Church of Christ Commission for Racial Justice. 1987. *Toxic Wastes and Race in the United States: A National Report on the Racial and Socioeconomic Characteristics of Communities with Hazardous Waste Sites*. New York.

U.S. Commission on Civil Rights. 2003. *Not in My Backyard: Executive Order 12898 and Title VI as Tools for Achieving Environmental Justice*. Washington, DC. *October.*

Vajjhala, Shalini P., Amanda Van Epps, and Sarah Szambelan. 2008. Integrating EJ into Federal Policies and Programs: Examining the Role of Regulatory Impact Analyses and Environmental Impact Statements. RFF Discussion Paper 08–45. Washington, DC: Resources for the Future.

Washington, Sylvia Hood. 2005. Packing Them In: *An Archaeology of Environmental Racism in Chicago, 1865–1954.* Lanham, MD: Lexington Books.

Waterman, Richard. 2004. Unilateral Politics. *Public Administration Review* 64 (2): 243–245.

West, Jonathan P., and Glenn Sussman. 1999. Implementation of Environmental Policy: The Chief Executive. In *The Environmental Presidency*, ed. Dennis L. Soden, 77–111. Albany, NY: State University of New York Press.

3

Federal Environmental Justice Policy in Permitting

Eileen Gauna[1]

Permitting is central to environmental regulation, and it is the primary means to attempt to curb the excesses of the Industrial Revolution that left residents of cities choking on blankets of smog, rivers so polluted that they caught fire, and land so contaminated that cancer clusters followed in their wake. While it is tempting to think that pollution should be banned altogether in light of all these problems, the realities of producing a range of necessary goods, from fabrics to electricity, counsel against such an absolutist view. The compromise is the permit, and the mechanism is the permit proceeding.

While authority for environmental permitting required under federal law often has been delegated to the states, who incorporate similar provisions in their own state laws, federal agencies still issue a fair number of permits. For example, various regions within the Environmental Protection Agency (EPA), in the absence of delegation to state permitting agencies, issue major source environmental permits under a range of federal environmental laws. In addition, the Federal Energy Regulatory Commission and the Nuclear Regulatory Commission issue permits for energy- and nuclear-related facilities, respectively.

To environmental justice advocates, the permit proceeding is critical because this is where it is determined for example, how much of an air pollutant can be emitted from a facility or how much toxic effluent will be allowed to flow from an outfall pipe into a nearby waterway. Advocates often see the permit as a gateway to a host of environmental inequities, such as nuisance impacts, lax enforcement, increased exposure to emissions, and risk of accidents and contamination. As such, permit applications spark intense controversies with high stakes for all sides.

Permit decisions can be administratively appealed, and the resulting published decisions demonstrate how federal environmental justice policy actually applies to permitting issues that arise in relatively large projects. In addition to these decisions, Agency guidance and planning documents articulate environmental justice policy in permitting, albeit in more general terms. A key example is the EPA's recently issued *Plan EJ 2014*, a purported roadmap for integrating environmental justice into the EPA's programs and policies going forward. *Plan EJ 2014* specifically contains a chapter on permitting.

Less direct, but still an important indicator of federal environmental justice policy in permitting, are the EPA investigations and the adjudication of claims under Title VI of the 1964 Civil Rights Act. These particular claims, a majority of which arose in the permitting context, allege that certain delegated state agencies use criteria or methods of administering their permit programs in a manner that subjects individuals to discrimination because of their race. These "disparate impact" claims—so termed because they need prove only discriminatory impact, not intent—have been particularly difficult for federal agencies to adjudicate.

The primary aim of this chapter is to evaluate the degree to which environmental justice considerations have become integrated into EPA permitting. It begins with a brief explanation of the permitting process in general and the Environmental Appeals Board (EAB) in particular. Next is a detailed examination of Board decisions on environmental justice challenges, providing an instructive lens into federal policy in this area. The earlier opinions were seminal decisions within the agency, ones where advocates questioned the permit issuer's adherence to a newly minted presidential executive order on environmental justice; these decisions set the framework for subsequent review. More recent opinions demonstrate the interplay between environmental justice challenges and health-based standards, giving a sense of federal environmental justice policy in the context of difficult permit decisions. The decisions also suggest how the Board's review itself shaped permitting policy by EPA regions. This chapter next examines how *Plan EJ 2014* purports to guide long-term environmental justice policy in permitting. Comparing *Plan EJ 2014* to the prior EAB administrative decisions can indicate whether the EPA's environmental justice policy, as articulated in the plan, may be charting a different course compared to recent past practices. Finally, this chapter briefly examines how federal environmental justice policy in permitting may bear on Title VI civil rights administrative claims pending before the EPA, as the fate of those claims sends important signals concerning the federal government's commitment to environmental justice.

The Permitting Process at the EPA

A permit proceeding, particularly one involving a large industrial facility, is a highly technical endeavor (e.g., EPA 1990). While a permit proceeding generally begins when a firm submits a permit application, even determining whether a permit is required can be complicated. Once the application is submitted, there is a review period where the EPA must determine whether additional information is necessary or whether the application is complete. Once the application is complete, the permitting official develops a draft permit that addresses how the applicable environmental standards will be met. This involves a number of judgments concerning the sufficiency of proposed pollution control technology, as well as the appropriate monitoring protocols and reporting mechanisms to support enforcement. Next, the EPA publishes the draft permit and opens a public comment period. During this time, interested persons or groups—such as environmental organizations, other agencies (state or federal), and community groups—can submit written comments raising concerns about permit terms. This may be the time when an environmental justice group will object to the siting or expansion of a facility in a heavily impacted area. There is often a public hearing on the permit. These procedures offer a key venue by which environmental justice advocates can intervene, through community-based organizations such as West Harlem Environmental Action, Inc. (WE ACT), alliances of environmental justice organizations such as the Indigenous Environmental Network, or public-interest law firms that represent environmental justice communities such as the Center on Race, Poverty, and the Environment. After this period, the permit may be revised to respond to comments and then finalized. The steps along the way might have fairly short time frames, such as a thirty-day comment period. By statute or rule, the agency might have a certain amount of time to make a final decision on the permit.[2] These time frames can make it challenging for groups to obtain an independent technical review of the permit provisions and adequately advance their interests, a practical limitation evident in the opinions discussed next.

The Environmental Appeals Board

The EAB is the final EPA decision maker on administrative appeals for permits issued under certain federal air, water, and hazardous waste law. The Board, created by William K. Reilly, the EPA administrator in 1992,[3] currently has four judges who are Senior Executive Service career agency attorneys (EPA 2013). While it is housed within the agency (physically outside the immediate Office of the Administrator), it is designed to be an impartial body, independent of other EPA functions. The Board typically sits in panels of three judges and makes decisions by majority vote. In practice, the Board is guided by its own prior precedent and standing orders, as well as its practice manual (Hazardous Waste Consultant 2013), although the latter is characterized as a guidance document, lacking the force of law (EPA 2013). The Board, comprised of senior attorneys (EPA 2014), are mindful of these legal nuances. Similarly, it cannot escape the Board's notice that federal environmental permitting statutes and regulations do not explicitly grant authority to consider environmental justice per se. And although legal observers have determined that such authority is implicit in various statutory provisions (Environmental Law Institute 2001; Lazarus and Tai 1999; National Academy of Public Administration 2001),[4] the scope of that authority is not precisely defined by statute or implementing regulations. The history of these agency adjudications is in large part a story of how this lack of clearly defined legal authority, combined with the extraordinary degree of deference the EAB gives to permit officials, present almost insurmountable barriers to successful permit challenges on environmental justice grounds.

The Environmental Appeals Board Cases

Over the past twenty years, the EAB has issued several important decisions that reveal the limited way in which environmental justice considerations have been incorporated into EPA permitting. This part of the chapter examines a set of early decisions that established the framework for environmental justice reviews in permitting, and then considers more recent decisions that may open the door for additional requirements in the area of health-based standards.

The Board Sets the Framework for Review

The Board has explicitly considered issues involving environmental justice concerns since 1993.[5] In the first two decisions, issued before former president Bill Clinton's Executive Order 12898 on Environmental Justice (EO 12898) in 1994, the Board very narrowly interpreted the scope of the EPA's authority to consider environmental justice and denied review (i.e., found no error in the permit proceeding).[6] Yet, one year after this executive order, the Board altered its approach. *In Re Chem Waste Management* (*Chem Waste*) involved a renewal of a hazardous waste facility permit in Fort Wayne, Indiana, which lies in EPA Region 5. In that case, the Board still denied review; however, this time the Board discussed environmental justice at length and emphasized the importance of considering the issue, albeit not to the degree that the challengers had requested. In *Chem Waste,* rather than requesting specific permit terms, the challengers appeared to argue that the permit for an expansion should be denied altogether based upon the mandate of EO 12898. The Board at this early date made it clear that the executive order would not operate to change substantive permit requirements; if regulatory requirements are met, the official must issue the permit regardless of the demographics of the community or economic effects of the facility on the community. Despite these limitations, however, the Board noted that "there are areas where the Region has discretion to act within the constraints of the Resource Conservation and Recovery Act (RCRA) regulations and, in such areas, as a matter of policy, the Region should exercise that discretion to implement the [EJ] Executive Order to the greatest extent practicable."[7] According to the Board, such discretion would encompass decisions about public participation and authority under an omnibus clause (a general clause covering items not specifically addressed under the statute in question).[8] The authority to deny a permit to protect health and environment would not depend upon finding a disparate impact on a minority or low-income community, but the existence of that omnibus clause authority would later become a key issue in civil rights challenges.

To environmental justice advocates, the importance of *Chem Waste* lies in its strong endorsement of what the Board termed "a more refined look" at environmental and health impacts within a community. An overly broad analysis of the community can mask a disparate impact on a particular part of it. In what would later become an often-cited portion of the decision, the Board noted that when a commenter submits a "superficially plausible" claim of disparate impact, the permitting official should include a health/environmental analysis focusing particularly on the environmental justice community suffering the claimed disproportionate threat. Thus, while the Board denied the relief requested, its seeming endorsement of the use of existing legal authority to protect vulnerable communities was encouraging.

But subsequently, the Board curbed the import of the broad language in *Chem Waste*. First, it limited the authority of the RCRA to consideration of health and environmental impacts only, specifically precluding economic impacts[9] or wider social ills. Second, and perhaps more important, the Board did not alter its highly deferential review. While Region 5 officials in *Chem Waste* had conducted an environmental justice analysis using a one-mile radius around the facility, the challengers submitted a study indicating that air pollutants from the facility could affect an African American community two miles away. In what would become a consistent pattern in environmental justice challenges, the Board afforded great deference to the "highly technical judgment" of the Region in defining the scope of the environmental justice analysis. The Board concluded that the challengers' study was too tentative to justify overcoming the deference normally afforded to the agency.

This early case set the framework for review of environmental justice claims, endorsing public participation and an environmental justice analysis but affording a high degree of deference to EPA permit writers. Given that the phrase *environmental justice community* and terms such as *disparity* lend themselves to varying interpretations, this deferential approach has often precluded successful environmental justice claims.[10] For example, the Board consistently considered the Region's choice of an area of impact—usually one or two miles surrounding the facility—to be one of those highly technical issues that warrant great deference.[11] This is an important limitation. In several cases, the Region's choice of radius was fatal to the advocates' claims.[12]

Such a high degree of deference to the agency appears inconsistent with *Chem Waste*'s direction to EPA permit officials to undertake a more refined look. For example, several later decisions suggest that compliance with a National Ambient Air Quality Standard (NAAQS) that was current at the time the permit was issued would preclude a finding of adverse impact. In such instances, however, the EPA does have the discretion to undertake a more refined look to make sure that there are no localized concentrations of pollutants (by requiring multisource modeling, for example). Despite this discretion, when an applicant's own single-source modeled air quality impacts had complied with minimal agency requirements,[13] the Board required the challenger to explain exactly why additional multisource analysis was necessary.[14] The logic of this view is that because a NAAQS is designed to protect human health while the air shed is in attainment, there can be no "adverse" effect of the type contemplated by EO 12898.[15] This approach is controversial. Advocates note, for example, that given the geographic expanse of an air shed and averaging of emissions within that region to determine NAAQS compliance, toxic hot spots may indeed occur, even if the air shed as a whole might be legally designated an attainment area under the Clean Air Act (CAA).[16]

Similarly, in the context of a RCRA permit, the Board placed a high burden on the environmental justice challengers to explain why an additional risk assessment might be necessary.[17] The overall effect of this approach is to leave the Board-endorsed "more refined look" solely in the hands of a permitting official lacking sensitivity to environmental justice concerns, or one otherwise hesitant to require additional protective measures.

Conversely, one scenario arises in which the Board will grant review and remand: when the permitting official summarily concludes that an environmental justice community is not affected without submitting into the record the details of a demographic analysis. For example, in an early case involving a Prevention of Significant Deterioration (PSD) permit, Region 9 permit officials relied on an employee memo concluding, without any supporting detail, that "after reviewing the project location and surrounding demographics . . . it was unlikely that an Environmental Justice issue applied."[18] The Board granted review on that issue and remanded, noting that such a cursory conclusion failed to show the basis for the Region's conclusion.

These early permitting cases illustrate that EPA permitting officials must conduct, and place into the record, an environmental justice analysis of some sort. If this is done, the permitting official can then rely upon that analysis to conclude either that the permit in question does not implicate an environmental justice community, or that because of compliance with a health-based standard, there is no adverse impact. More recent Board cases might call that conventional wisdom into question somewhat, although it is too early to tell whether there has been a significant shift in this framework of Board review.

Mixed Signals Concerning Health-Based Standards

In one recent case, the Board arguably applied a more searching review and remanded for failure to perform an acceptable environmental justice analysis. Moreover, the Board was in fact willing to look beyond the currently enforced legal NAAQS designation of attainment. *In Re Shell Gulf of Mexico, Inc., Shell Offshore, Inc. (Frontier Discovery Drilling Unit) (Shell II),*[19] was one of several related cases involving Shell's exploration in the Chukchi and Beaufort Seas with large anchored drilling vessels and associated fleets of support ships, such as ice breakers and supply ships. The operation was anticipated to result in a substantial amount of air emissions. Such air pollutants could potentially impact traditional subsistence food sources and the low-income northern slope Inupiat communities. They are an indigenous population that the EPA associates with vulnerability or susceptibility, and who appear to have a relatively higher incidence of upper respiratory problems and diabetes.[20] Region 9 concluded that permitted activities could potentially impact these environmental justice communities. However, the Region relied solely on compliance with the NAAQS for nitrogen dioxide (NO_2) that was in effect at the time the permits were issued. The Region attempted to persuade the Board that because the permits would not cause exceedance of that standard, the North Slope communities would not suffer a disproportionately high and adverse human health effect from issuance of the two PSD permits. However, several months before the permits had been issued, the EPA administrator published a proposed rule in the Federal Register that concluded, based on updated scientific information, that the annual NO_2 NAAQS no longer adequately protected public health. She therefore established a supplemental 1-hour NO_2 NAAQS. There was a timing complication, however. At the time the permits had been issued to Shell, the new standard had been finalized but had not yet gone into effect. Accordingly, the Region argued that when the permit was issued, the area was legally designated in attainment with the then-applicable annual NAAQS for NO_2.

The Board noted that it had accepted such arguments in earlier decisions, but in those cases, the then-effective standard had not been called into question by the regulatory agency itself. Moreover, the environmental justice challengers did not have to submit detailed evidence to rebut the Region's argument that a NAAQS is presumptively protective. A collateral rule-making record contained ample epidemiologic evidence questioning the sufficiency of the then-current annual NO_2 standard. The Board noted that this scientific evidence was available to the public and the Region, through the EPA's electronic docket, more than eight months prior to the issuance of either permit.

This victory for the environmental justice challengers was short-lived. On appeal after remand, the Board endorsed the Region's supplemental twenty-page environmental justice analysis. Again noting deferential review of technical issues, the Board accepted the Region's conclusion that NO_2 values, according to new modeling—modeling that addressed air quality 500 to 2,000 meters from the hull of the vessel—would not exceed the *new* hourly NO_2 standard.[21]

Although the environmental justice challengers ultimately failed, the case held out the possibility that a challenge could succeed notwithstanding compliance with a health-based standard, if the standard itself were shown to be insufficiently protective. The case also held out the possibility that the Board could be persuaded to undertake more searching review of the adequacy of the permitting agency's environmental justice analysis. Subsequently, however, the *Shell II* opinion proved to be of more limited import than one might have anticipated.

After the new hourly NO_2 standard took effect, a similar challenge arose to the issuance of another PSD permit for a new 600-megawatt (MW) natural gas–fired power plant in Avenal, California. *In Re Avenal Power Center, LLC (Avenal)*, involved a complaint, submitted to the EPA by Greenaction for Health and Environmental Justice, alleging that the district discriminated against residents of color and Spanish-speakers in the city of Avenal and the nearby town of Kettleman City by failing to notify or involve them during the decision-making process.[22] Much of the area is agricultural and annually hosts a large number of migrant laborers. Greenaction also alleged that these residents were already impacted by multiple sources of pollution and that the operation of the proposed Avenal power plant would result in adverse health impacts. The permit terms, according to data submitted by the challengers, would result in emissions violating the new NO_2 NAAQS.[23] Here, the EPA had prepared a thirty-page environmental justice analysis. The analysis used monitoring data from EPA-approved sites located 28 and 46 miles from the proposed facility, among other data, to assess the short-term exposure of communities of poor people and people of color to NO_2 emissions from the facility. Based on this limited data, the EPA determined that it could not reach a definitive conclusion about specific human or environmental impacts. In this case, the agency also determined that it would be appropriate to "grandfather" the facility (i.e., allow it to use the prior, less protective standard), because a court had imposed a strict deadline for the agency to make a decision on issuing the permit. Thus, because of a prior successful procedural challenge of agency delay—which by itself illustrates a potentially significant barrier to an environmental justice challenge[24]—and lack of data to show whether the new standard had been met, the more stringent NAAQS was not helpful to the environmental justice challengers.

While those circumstances might seem anomalous, a closer look at the NO_2 issue in the case illustrates the difficulty that environmental justice challengers have in establishing a claim that will afford them a remedy. Despite the decision to grandfather, the agency nonetheless discussed NO_2 hourly impacts in its environmental justice analysis, presumably because of the high rates of respiratory illness in the area and the localized effect of NO_2.[25] And while the data were not optimal, the available information arguably raised legitimate environmental justice concerns. First, the EPA had concluded that limited data "available from the [monitors] located closest to the proposed facility indicate that background levels of hourly NO_2 'are on par with measured levels of NO_2 statewide,' and thus background levels of hourly NO_2 in the general area of the facility are not disproportionately high as compared to communities throughout the state." However, the environmental justice analysis also disclosed that operational emissions from the Avenal facility alone would represent

44 percent of the standard—and further acknowledged the absence of detailed information about how those emissions might combine with nearby and background sources to affect compliance with the hourly NO_2 standard. Moreover, while a facility several miles away from the Avenal facility provided NOx offsets, those offsets did not mitigate the localized impacts of NO_2.

In addition, the environmental justice challengers submitted information suggesting that the hourly NO_2 standard would likely be violated, particularly since hourly NO_2 emissions in the nearby town of Kettleman City, located directly adjacent to Interstate 5, would necessarily be higher than emissions at the monitoring stations farther away. Yet, the Board's response to this concern was simply to note the considerable leeway that the EPA has in complying with EO 12898 and that petitioners challenging the adequacy of an environmental justice analysis bear a "particularly heavy" burden in overcoming deference to the agency's technical expertise.[26] The Board also observed that its previously issued *Shell II* decision did not require the agency to reach a determinative conclusion about NO_2 compliance prior to issuing the permit. The Board distinguished the remand in *Shell II* because it had turned on the Region's "scant environmental justice analysis, which provided no examination or analysis of short-term NO_2 impacts whatsoever." In a petition for review of the Avenal matter, the environmental justice challengers noted the irony that the EPA lacked information about disparate impacts only because the agency itself grandfathered the project from any need to demonstrate impacts and compliance.[27]

In addition to limiting the potential reach of *Shell II's* inquiry into the sufficiency of a NAAQS to protect the community at issue, the Board in *Avenal* potentially signaled an additional barrier to successful environmental justice claims. In discussing the adequacy of the Region's cumulative impact analysis, the Board endorsed the EPA's approach of addressing cumulative impacts through mitigation strategies outside the permitting process, such as enforcement actions, ongoing local air quality planning processes and administrative compliance orders under the Safe Drinking Water Act (SDWA).[28] Finally, the Board noted approvingly that the EPA was then conducting an agencywide strategic planning process to address environmental justice, particularly the complex issue of cumulative impacts.

Subsequently, in an appeal following the remand of *Shell II*,[29] environmental justice challengers again attempted to use the EPA administrator's own rule-making record against the agency's decision to issue the permit. This time, the challengers argued against the sufficiency of the current ozone NAAQS by pointing to the administrator's proposal for a new, more stringent 8-hour ozone standard. With this argument, apparently modeled after the successful argument in *Shell II*, the environmental justice challengers would again have the benefit of a collateral rule-making record that contained evidence of the insufficiency of the current standard. However, the argument failed this time. The Board observed that four days before the close of the public comment period and prior to the Region issuing the permit, President Barack Obama had requested that the EPA administrator withdraw the proposed standard, continue to enforce the current standard, and reconsider the standard again in 2013. The Board's reliance upon the current legal standard, in light of the rationale of *Shell II*, is curious in at least one respect. The health-related data in the ozone rulemaking was not directly at issue in the president's decision, which media reports attributed largely to his concern with the economic and political effect of the new standard (Broder 2011). The Board did observe, however, that the Region had exercised its technical expertise to determine that the ozone levels in the area were not expected to exceed even the lowest proposed standard. What is not clear is how much weight the Board gave the latter determination, as it then cited the generally applicable principle that a permit issuer must apply statutes and implement regulations in effect at the time the final permit is made.

In theory at least, these cases together suggest that the Board could conceivably uphold a denial of a permit based on environmental justice considerations under its often-applied principle of strong agency deference, assuming the Board would determine that there was adequate legal authority under an applicable omnibus clause. However, the Board's tendency to subsequently limit the reach of prior case reasoning favorable to environmental justice challengers makes such a prediction risky. Such a case would have to be extremely compelling. Moreover, the more recent *Shell* decisions, together with *Avenal*, leave unclear the question of whether the Board is relying primarily upon the health-based standard that is legally in effect at the time of the permit issuance, or instead will evaluate the sufficiency of the standard for protecting health, notwithstanding the current legal designation. Again, this is an important issue for environmental justice challengers, who often assert that because of the general insufficiency of standards, localized concentrations, or extraordinary health vulnerabilities in the community, reliance on NAAQS or other health-based standards is not always appropriate. (Chapters 4 and 5 discuss this issue in more detail.) It also appears that despite EO 12898's directive to be protective, lack of data is likely to favor the permit applicant over the vulnerable community, primarily because of deference to the technical judgment of the agency. This approach appears to be inconsistent with a directive to undertake a more refined look for potential impacts. In addition, both *Avenal* and *Shell II* analyzed the effect of emissions on the NAAQS[30] while underemphasizing the impacts on the communities themselves from emissions and broader facility operations.[31] Finally, where a disparate impact is shown, offsite mitigation might be used in some as-yet-undetermined manner to offset the facility's disproportionate impact on the vulnerable community.

Collectively, these cases illustrate that permit officials in the first instance are hesitant to use their legal authority to impose significant additional permit conditions, much less to deny a permit altogether due to environmental justice considerations. Instead, permit officials appear more willing to make technical determinations that avoid this problematic legal territory. The EAB in turn has avoided similar inquiry by the application of a deferential review, one that is fairly high for an internal agency review (Shelby 2013). One may speculate that a more searching review by the Board, examining whether permit officials had actually undertaken a more refined look at potential environmental justice impacts, would likely have resulted in more protective environmental justice policy in permitting.

Apart from the high burden placed upon environmental justice challengers, there is one aspect of recent federal permit review developments that might indirectly benefit vulnerable communities. Because of the Board's willingness to press the issue, permitting agencies now appear inclined to submit a better developed environmental justice analysis into the record than they had previously. The environmental justice analyses in the *Shell II* and *Avenal* permit proceedings discussed demographic characteristics, cultural practices, and health vulnerabilities and disparities.

For example, population, age, and race were charted.[32] While this is a commendable effort, screening approaches of this nature should be used with caution. A National Environmental Justice Advisory Council (NEJAC) workgroup cautioned that the use of particular demographic factors, along with weighting of those factors, can skew the assessment of impact (National Environmental Justice Advisory Council 2010).[33] All told, however, whether regional permitting officials will eventually redirect and refine their analysis to consider more localized cumulative impacts and health vulnerabilities, as well as a potentially broader range of impacts on communities, is still unclear. However, high-level agency policy statements have recently seemed inclined in this direction, thus potentially prompting permitting officials toward more refined analysis and promoting technical tools and agency legal interpretations to support more protective permitting measures. Environmental justice advocates have turned their attention to such indications.

Plan EJ 2014: Considering Environmental Justice in Permitting

In September 2011, then–EPA administrator Lisa Jackson, anticipating the twentieth anniversary of EO 12898 in February 2014, issued *Plan EJ 2014,* billed as a comprehensive environmental justice strategy (EPA 2011). The plan focused on critical environmental justice areas, including rulemaking, permitting, compliance, and enforcement, as well as community-based programs and working with other federal agencies. On permitting in particular, the strategy's goal is to give environmental justice communities meaningful access to the permitting process, while encouraging the EPA to develop permits that address environmental justice concerns to the extent practicable under existing laws. As early as July 2010, the EPA took comments on a draft of this plan, responded to those comments, and later incorporated some suggestions (EPA, no date). In April 2012, the EPA issued a draft supplement, specifically addressing Title VI of the Civil Rights Act (EPA 2012), as well as a tribal consultation and coordinating plan (EPA, no date). Despite some criticism that *Plan EJ 2014* documents are unduly vague, they are perhaps the most comprehensive statement of official agency policy on environmental justice to date, and the EPA has committed to make annual reports of its progress in achieving its goals.[34]

In the chapter on permitting in *Plan EJ 2014*, the agency also articulated a commitment to creating a culture within the EPA, as well as other federal, state, local, and tribal permitting venues, that translates engagement on environmental justice issues into greater protections for overburdened communities. The chapter envisioned a set of environmental justice regulatory "tools," defined loosely to include guidances, best practices and checklists (e.g., for public involvement and communication), templates, reports, case studies, mapping and screening instruments, protocols, trainings, sample languages, and social networking resources (e.g., communication between permit writers and stakeholders via postings and RSS feeds). The EPA would solicit recommendations to develop and test these tools. The strategy focused on EPA-issued permits in the short term, but with an eye toward all permits issued under federal law (EPA 2011). The plan's appendix also listed specific agency activities, such as literature reviews, workgroups, case studies, and implementation plans.

Significantly, the EPA also noted that it could develop guidances on conducting environmental justice analyses (including integrating environmental justice into permit conditions and mitigation actions), use of screening tools, cumulative impact methodologies, and exposure-based (health-effects) modeling. Of particular note was the development of guidances and trainings on addressing environmental justice concerns through provisions concerning the best available control technology, offsets, monitoring, record keeping and reporting, startups, shutdowns and malfunctions, lower potential to emit, emission factors, and CAA Title V operating permits. This is meaningful, as outside reviews, policy papers and reports have long encouraged the EPA to use existing statutory and regulatory provisions to develop guidances on environmental justice (Environmental Law Institute 2001; National Academy of Public Administration 2001, 2002), particularly focusing on provisions of the CAA (Gauna 1996; National Academy of Public Administration 2003). In this respect, the EPA's strategy is ambitious.

As ambitious as the plan is, however, many who responded via comments felt that it lacked essential details concerning implementation, progress measurement, and mechanisms for accountability. With respect to permitting in particular, several commenters noted the importance of early and robust public participation, even at the preapplication stage. Others noted, however, that "[a]ttempting to force too early of a disclosure may get communities worked up for no reason...." Addressing such issues, the EPA observed that, while it will focus initially on EPA-issued permits, state permitting authorities wield substantial expertise of their own. Hence, *Plan EJ 2014* also will collect best practices by the states and compile them into a resource compendium. Some of these best practices include mandatory public participation plans as part of permit applications. The agency's ringing endorsement of public participation and an environmental justice analysis in permitting fairly tracks the EAB's encouragement of these practices, as well as the relatively more comprehensive environmental justice analysis referenced in later Board opinions (EPA, no date).

However, in one additional area, *Plan EJ 2014* seems to go beyond the Regions' current practices, and also beyond what the Board has been willing to prompt the EPA to do: evaluate cumulative risks and impacts. One commenter called on the EPA to develop methodology for such analysis that regards "both large and small facilities, including those that would otherwise not be permitted or regulated, and vehicles." Another commenter, however, felt that "such an analysis, while laudable in its intention, [would be] impractical in anything more than a qualitative sense." According to this commenter, current science could address such impacts only "via a multi-media cross program implementation approach." Such impacts, it was noted, could not be addressed simply through the EPA's policy pronouncements. The agency delivered a lengthy response, giving three detailed reasons from EPA's Risk Assessment Forum that not only supported the feasibility of cumulative risk assessment on environmental justice at this time, but also anticipated increasingly robust methodologies in the future. The response also referenced draft guidelines for cumulative risk analysis that the agency was finalizing. The EPA concluded:

[T]he approach being developed at EPA to enable [cumulative risk analysis] employs a range of information, from qualitative to quantitative; integrates cutting-edge science to understand disease pathways and how they are activated by stressors; and uses planning and scoping approaches to narrow the scope of [cumulative risk analyses] so that the analysis is manageable but also focused on the most important factors that may affect human and environmental health (EPA, no date).

On another aspect of cumulative impacts related to permitting, the agency agreed that the "[EPA] must provide guidance on how to address the existing concentration of polluting facilities in low-income communities and communities of color" (EPA, no date).

These relatively strong statements on permitting from *Plan EJ 2014* appear at odds with the *Avenal* decision, where the Region undertook only limited analysis of cumulative impacts, as well as earlier EAB decisions where the Regions did not perform multisource modeling and other types of analyses relevant to assessing cumulative impacts.[35] The EPA's response to comments on these issues also departs from the agency's long-standing hesitancy to develop guidance on addressing environmental justice under potentially applicable provisions of the environmental statutes and implementing regulations. This slight but significant shift in position is likely due in part to continued pressure from environmental justice communities. In addition, the methodologies in environmental justice empirical research and risk assessment have advanced, although the quality of local land use and health data continues to vary significantly.

While it is unlikely at this stage that permitting authorities will deny a permit solely on environmental justice grounds (absent some noncompliance with specific permitting requirements), a more robust cumulative impact analysis could, in time, prompt more aggressive mitigation. Providing technical guidance on conducting a cumulative impact analysis in a permit proceeding will be challenging but feasible, as shown by similar, existing draft EPA technical guidance on environmental justice in rulemaking proceedings (EPA 2013c). Coupled with the interest in on-site and off-site mitigation, these developments hold some promise of moving beyond enhanced participation in permitting to actually reducing impacts (or at least stemming the growth in adverse impacts) in disparately impacted communities. Historically, the agency's failure to aggressively address cumulative impacts in permit proceedings has made improving conditions in environmental justice communities elusive. If the EPA moves forward assertively in this respect at the regional level, the agency can foster more protective permitting not only with respect to EPA-issued permits, but in states exercising delegated federal permitting authority under formal agreements with the EPA. While resistance from state permitting agencies and business interests is predictable, the agency will nonetheless be better positioned to demand more serious state environmental justice action under environmental statutes and regulations. In addition to environmental laws, the substantive nondiscrimination mandate of Title VI, a civil rights law, can significantly affect permitting at the state level.

Federal Environmental Justice Policy as Reflected in EPA's Title VI Enforcement Efforts

Title VI of the Civil Rights Act of 1964 prohibits racial discrimination by recipients of federal funding. The U.S. Supreme Court interpreted the statute to authorize federal regulations precluding recipients from activities having a discriminatory effect,[36] rather than prohibiting only intentional discrimination. The EPA first promulgated regulations aimed at alleviating discriminatory effects in 1973.[37] These regulations currently provide that "[a] recipient shall not use criteria or methods of administering its program which have the effect of subjecting individuals to discrimination because of their race."[38] In 1993, environmental justice advocates began using the administrative procedures under these regulations to file complaints, many of them against state permitting officials, whom advocates alleged created or exacerbated racial disparities by continuing to issue permits allowing releases of pollutants in heavily impacted communities of color. The ensuing saga of Title VI at the EPA involved a controversial interim guidance, a federal advisory group unable to agree on a comprehensive set of recommendations, congressional hearings, riders on bills prohibiting the EPA from using appropriated funds to enforce Title VI, a set of never-finalized draft guidances issued in 2000, and hundreds of complaints that were either dismissed on technical grounds or left to languish at the agency for years (Deloitte Consulting 2011; EPA, Civil Rights Executive Committee 2012).[39] While the intricacies of Title VI adjudication and the truly difficult set of issues that arise in the permitting context are beyond the scope of this chapter (see chapter 8 for additional discussion), federal environmental policy concerning the role of environmental standards is reflected in recent Title VI initiatives, which are briefly discussed next.

After more than a decade of apparent inaction, the admonishment of the EPA by at least one federal court[40] and the continual urging of advocates, the agency began to take action. In 2012, it issued a supplement to *Plan EJ 2014* specifically addressing Title VI (EPA 2012). In that year, the agency also resolved some long-standing complaints. In one instance—a matter not involving the issuance of a facility permit—the EPA in the case known as *Angelita C* issued its first formal preliminary finding of a Title VI violation prior to a settlement of the matter.[41] In another case (one also not involving a permit), the agency entered into a settlement to resolve a complaint alleging that a state agency's in-person interview requirement for investigating violations of worker protection standards created a disparate impact based on national origin.[42] In a third case, one that did involve a permit, the EPA had partially accepted for investigation the 1994 *Padres* complaint, alleging that several state and county agencies had discriminated against Latinos in the siting, permitting, and public participation processes for California's three commercial hazardous waste disposal facilities.[43] The EPA ultimately found no violation, dismissing the *Padres* complaint altogether. Finally, in a related development, the agency issued in January 2013 what it termed a draft "white paper" on the adversity element of a Title VI claim, addressing the contentious issue of the effect of compliance with health-based standards (EPA 2013d).

Environmental justice advocates found much that was troubling in this flurry of activity. The *Padres* decision was an obvious disappointment, and as for the settlement of the *Angelita C* case, advocates bitterly complained that the remedy (more monitoring) was completely inadequate.[44] They noted that "the California Department of Pesticide Regulations ... was not in jeopardy of losing funds provided by the EPA for the application of the toxic pesticide methyl bromide on Latino schoolchildren and DPR was not held accountable for its actions." Environmental justice advocates also criticized the lack of opportunity for communities to participate in Title VI settlement negotiations.

As for the Title VI Supplement to *Plan EJ 2014*, while they stated that they appreciated the effort, advocates noted the general lack of specificity concerning implementation. With respect to the ongoing issue of the interplay between Title VI and federal environmental laws, they noted:

Compliance with environmental laws and standards should not be the ruler for civil rights compliance. Title VI is a civil rights statute, and it is independent of environmental laws and standards. ... [T]hese standards may involve averaging emissions over large geographical areas that, if viewed in isolation, can hide disparities. They are, again, not the benchmark for a determination of "impact." Among other things, environmental standards do not fully capture harms to public health and the environment. These standards change over time, for instance, precisely because they are found to be insufficiently protective.[45]

Environmental justice advocates and others are keenly interested in how the EPA may resolve Title VI complaints, not only because state agencies issue the majority of environmental permits, but also because of the implications for federally issued permits. To the latter point, it would be politically difficult for the EPA to interpret the states' nondiscrimination mandate of Title VI more stringently than it would interpret its own duties under EO 12898. To be sure, the executive order and *Plan EJ 2014* are essentially statements of administrative policy, while Title VI and its implementing regulations have the force of law. Yet, the federal government should be willing to reach the same level of protection under the executive order and the plan as it requires state permitting officials to do under Title VI. Given the barrier that NAAQS compliance presented to environmental justice advocates before the EAB, as well as in at least one high-profile Title VI adjudication,[46] if the EPA were to abandon its inclination to rely solely on technical compliance with health-based standards and instead require a more rigorous analysis of localized and cumulative impacts, this shift could offer greater protection for impacted communities.

To a limited degree, this might be in fact occurring, at least within some quarters of the EPA. In the white paper on Title VI, the agency noted its reconsideration of a long-standing view, first articulated in a set of draft guidances issued in 2000, that compliance with a health-based standard would raise a "presumption" in a Title VI investigation against a finding of adverse impact to the community.[47] Perhaps one of the most interesting aspects of this white paper is the agency's observation that

this proposal acknowledges the relative significance of compliance with an environmental health-based threshold, while also evaluating a number of other factors, as appropriate, including the existence of hot spots, cumulative impacts, the presence of particularly sensitive populations that were not considered in the establishment of the health-based standard, misapplication of environmental standards, or the existence of site-specific data demonstrating an adverse impact despite compliance with the health-based threshold. ... (EPA 2013d)

In principle, this observation seems to capture what the EAB likely meant in *Chem Waste*'s early reference to a more refined look, as well as what the EPA was addressing in *Plan EJ 2014* and its subsequent response to comments. In a related vein, the EPA reexamined in another draft white paper its view of the role of complainants in Title VI complaint investigations and settlements (EPA 2013e). While the agency affirmed its prior position denying a formal role for the complainant in a

Title VI investigation, it did indicate its intent, in appropriate instances, to afford complainants an opportunity to participate in alternative dispute resolution. Also, in some instances, the EPA agreed to allow complainant input regarding appropriate remedies following a preliminary finding of noncompliance. These informal steps toward greater inclusion indicate (albeit to a limited degree) an intent to depart from the status quo.

Concluding Thoughts

After the signing of EO 12898, the EAB encouraged stronger public participation and endorsed environmental justice analyses in permit proceedings, but because of its highly deferential review, it did little else to inquire about or prompt more protective measures to address environmental justice concerns. While deference to permit officials' technical expertise is normally appropriate, the Board can undertake a more probing inquiry. Even in situations demanding deferential review, federal courts at times apply a form of "hard look" judicial review that seeks to determine whether agency officials carefully review the salient issues in question. Here, EO 12898 and the EPA's commitment in *Plan EJ 2014* to promote a permitting culture that translates engagement on environmental justice into greater protection for vulnerable communities both warrant a similar type of inquiry.

It is likely that much of the hesitancy to exercise discretion more aggressively and use interpretive authority—either at the permit writing level or during internal agency review of permit decisions—stemmed from a lack of specific legal authority and regulatory tools to address these complicated issues. Upon the twentieth anniversary of EO 12898, the EPA's acknowledgment that a more refined analysis of cumulative impact and risk is feasible, its interest in regulatory tools to further this important goal, and its aspirations stated in *Plan EJ 2014* are cause for cautious optimism. Environmental justice will benefit if these policy directives are in fact implemented at the permit level within the EPA and applied to states indirectly in Title VI investigations.

Unfortunately, the permitting venue may be particularly resistant to addressing a variety of environmental justice concerns, not only because of opposing political and economic pressures, but also because, in this highly technical arena, participants rely heavily on statutes and regulations that clearly define permit requirements. As a practical matter, for permitting officials to be more willing to impose additional requirements or deny a permit, the EPA must first clarify its legal authorities and provide more specific guidance concerning measures that can be taken to address environmental justice considerations. It is noteworthy that in the past, the EPA has creatively interpreted legal authority at the behest of industry, leading EPA staff to famously characterize a Clinton-era regulatory reinvention initiative (Project XL): "if it isn't illegal, it isn't XL" (Steinzor 1996). More recently, the EPA has been willing to interpret its legal authority to regulate greenhouse gases in a manner that industry views as overly expansive.[48] The lesson here is that agency willingness to use its interpretive authority—or not—can itself have environmental justice implications. The predictable fallout of state and industry challenges will necessarily test the EPA's political will to keep the promises in *Plan EJ 2014*.

Apart from agency initiatives, academic environmental justice research on measuring impacts and risk continues to advance. Additional studies might empirically assess how the EPA and state permitting agencies in fact exercise case by case discretion in permit proceedings that implicate environmental justice scenarios. The findings could shed light on long-standing concerns that permit officials are too timid in using their discretion to protect vulnerable communities.

In summary, the intentions expressed in more recent EPA directives, our increased understanding of and ability to measure cumulative impacts, and the EPA's willingness to develop regulatory tools all offer encouraging signs that federal environmental justice policy may advance toward greater protectiveness in permitting. An important next step will be the EPA's willingness to interpret its legal authority and give guidance on specific measures to provide that protection.

Notes

1. Special thanks to my research assistant, Ed Merta.

2. See, e.g., James Pierce Marlborough Wiegard, "Finding a Dog That Will Hunt: Solutions for Interagency Permit Consultation Delays Related to Prevention of Significant Deterioration Permits in the Wake of *Avenal Power Center v EPA*," 14 Vt. J. Envtl. L. 635 (2013), which discusses how interagency consultations can present problems to the permitting agency working under tight deadlines to issue permits (hereinafter *Interagency Permit Consultation Delays*).

3. *Changes to Regulations to Reflect the Role of the New Environmental Appeals Board in Agency Adjudications,* 57 Fed. Reg. 5320 (Feb. 13, 1992).

4. See also Memorandum from Gary Guzy, General Counsel, to the assistant administrators of various program offices within the EPA (Dec. 1, 2000).

5. For a discussion of the first eight EAB cases that raised environmental justice challenges, see Foster (2000).

6. Genesee Power Station Ltd. P'ship, 4 E.A.D 832 (EAB 1993), 1993 WL 484880; opinion reissued to respond to a motion for clarification, 1993 WL473846, available at http://yosemite.epa.gov/oa/EAB_Web_Docket.nsf/e9b915d28d695a7785256e-8400458b9e/593dc42f06cc64ab85257069005f7ddf!OpenDocument. (Unless otherwise noted, all of the EAB decisions cited hereinafter are available at http://yosemite.epa.gov/oa/EAB_Web_Docket.nsf/Case~Name?OpenView, listed in alphabetical order, with the decision date and reporter citation accompanying the listing for each case.)

7. *Chem. Waste Mgmt., Inc.,* 6 E.A.D. 66, 77 (EAB 1995), 1995 WL 395962.

8. *Id.* at 74. (citation omitted).

9. See, e.g., Upper Blackstone Water Pollution Abatement Dist., NPDES Permit No. MA 0102369, slip op. at 104–105 (E.A.B. May 28, 2010), 2010 W.L 2363514.

10. For example, EPA Region 10 noted that the extent of an environmental justice analysis will vary according to the unique circumstances of each case. Envl Prot. Agency Region 10, Supplemental Environmental Justice Analysis for Proposed Outer Continental Shelf PSD Permit No. R10OCS/PSD-AK-2010 & Permit No. R10OCS/PSD-AK-09–01at 1 (undated) [hereinafter *Shell Supplemental Environmental Justice Analysis*], http://yosemite.epa.gov/oa/eab_web_docket.nsf/Filings%20By%20Appeal%20Number/4BB1D10E49B-2C0F585257934006FEFB8/$File/Final%20Attachment%204...11.pdf.

11. See, e.g., Envtl. Disposal Sys., Inc., 8 E.A.D. 23, 35–36 (EAB 1998), WL 723912; Beeland Grp., LLLC, Beeland Disposal Well #1 UIC Permit No. MI-009–1I-001, slip op. at 26–28 (EAB 2008), 2008 WL 4517160.

12. *See, e.g.,* Chem. Waste Mgmt, at 75–81 (1-mile radius); Envotech, L.P., 6 E.A.D. 260, 277–283 (1996), 1996 WL 66307 (within 2-mile radius); Envtl. Disposal Sys., Inc., 8 E.A.D. 23, 36 (EAB 1998), 1998 WL 723912 (within 2-mile radius).

13. See, e.g., AES Puerto Rico, 8 E.A.D. 324, 331 (EAB 1999), 1999 WL 345288; EcoElectrica, L.P., 7 E.A.D. 56, 65–66 (EAB 1997), 1997 WL 160751.

14. EcoElectrica, 7 E.A.D. 56 at 67; Phelps Dodge Corp., Verda Valley Ranch Dev., 10 E.A.D. 460, 461–462 (EAB 2002), 2002 WL 1315601.

15. Shell Offshore, Inc., Kulluk Drilling Unit and Frontier Discoverer Drilling Unit, 13 E.A.D. 357, 405 (EAB 2007), 2007 WL 3138040.

16. Letter from environmental justice advocates to EPA Administrator Lisa Jackson (July 3, 2012) (on file with author).

17. Ash Grove Cement Co., 7 E.A.D. 387, 413 (EAB 1996), 1996 WL 732000.

18. Knauf Fiber Glass, GMBH, 8 E.A.D. 121, 175 (EAB 1999), 1999 WL 64235.

19. OCS Permit No. R10OCS/PSD-AK-09–01 & OCS Permit No. RL0OCS/PSD-AK-2010–01, slip op. (E.A.B. 2010), 2010 WL 5478647.

20. *Shell Supplemental Environmental Justice Analysis, supra* note 10, at 5–12.

21. Shell Gulf of Mexico, Inc., OCS Permit No. R10OCS/PSD-AK-09–01 & Shell Offshore, Inc., OCS Permit No. R10OCS/PSD-AK-2010–01 Noble Discoverer Drillship, slip op. at 37 (E.A.B. 2012), 2012 WL 119962.

22. See chapter 6 for a discussion of public participation issues.

23. Avenal Power Ctr, LLC, Permit No. SJ 08–01, slip op. at 14–18 (E.A.B. 2011), 2011 WL 4881823.

24. *Cf., Interagency Permit Consultation Delays, supra* note 2; see also Gary McCutchen and Colin Campbell, "Horsefeathers!": Landmark Court Decision Directs EPA to Address Grandfathering, CAA One-Year Permit Processing Mandate," 21 No. 4 *Air Pollution Consultant* 5.1 (2011).

25. EPA Region 9, Supplemental Statement of Basis PSD Permit Application for Avenal Energy Project 20 (March 2011) [hereinafter *Avenal Supplemental Statement of Basis*], http://www.epa.gov/region9/air/permit/avenal/Avenal-Supp-StatemtBasisEjAnalysisApdxFinal-Eng3-2-11.pdf.

26. Avenal, Permit No. SJ 08–01, slip op. at 26–27.

27. Earthjustice Petition for Review at 38, Avenal Power Center, LLC, Permit No. SJ 08–01 (E.A.B. 2011) (undated), http://earthjustice.org/sites/default/files/Avenal-Petition-for-Review.pdf.

28. Avenal, Permit No. SJ 08–01, slip op. at 27–28.

29. Shell Offshore, Inc., OCS Permit No. R10 OCS03, slip op. (E.A.B. 2012), 2012 WL 1123876.

30. With the exception of *Avenal's* reliance on lack of data on hourly NO_2. See Avenal, Permit No. SJ 08–01, slip op. at 24–27.

31. See, e.g., EarthJustice Petition for Review, *supra* note 27 at 35–39.

32. *Shell Supplemental Environmental Justice Analysis, supra* note 10, at 9–12; *Avenal Supplemental Statement of Basis, supra* note 25, at 17, 22, 25.

33. See also McArdle (2012), noting the limitations of EPA's subsequently developed mapping tool, EJSCREEN.

34. As of March 2013, EPA has produced one such progress report; see EPA (2013b).

35. See, e.g., AES Puerto Rico L.P., 8 E.A.D. 324, 331, 335–341, 350–352 (EAB 1999), WL 345288.

36. *Guardians Ass'n v. Civil Serv. Comm'n of the City of N.Y.,* 463 U.S. 582, 584, 593 (1983).

37. *Nondiscrimination in Programs Receiving Federal Assistance from the Environmental Protection Agency,* 38 Fed. Reg. 17,968, 17,969 (July 5, 1973); see also *Non-Discrimination on the Basis of Race, Color, National Origin, Age,*

Handicap, and Sex in Federally Assisted Programs, 46 Fed. Reg. 2306 (January 8, 1981); *Nondiscrimination in Programs Receiving Federal Assistance from the Environmental Protection Agency,* 49 Fed. Reg. 1656 (Jan. 12, 1984), codified at 40 C.F.R. pt. 7 and 12.

38. *Nondiscrimination in Programs or Activities Receiving Federal Assistance from the Environmental Protection Agency,* 40 C.F.R. pt. 7.35(b) (2000). See also 40 C.F.R. pt. 7.35(c) (2000).

39. For a discussion of the history of Title VI claims at the EPA, *see* LoPresti (2013), noting 247 complaints as of 2013, and Eileen Gauna (2001) examining draft guidances.

40. *Rosemere Neighborhood Ass'n v. U.S. Environmental Protection Agency,* 581 F.3d 1169 (9th Cir. 2009).

41. See *Angelita C. v. California Department of Pesticide Regulation,* Title VI Complaint No. 16R-99-R9, EPA, Office of Civil Rights; Preliminary Finding, Title VI Complaint 16R-99-R9, EPA, Office of Civil Rights, (Apr. 22, 2011); *Investigative Report for Title VI Administrative Complaint File No. 16R-99-R9,* EPA, Office of Civil Rights, (Aug. 25, 2011). These and other materials related to the complaint are available at http://www.epa.gov/civilrights/TitleVIcases/index.html#angelita. For a discussion of the case, *see* LoPresti (2013) at 790–793.

42. Settlement Agreement for Administrative Complaint No. 04R-08-R6 and Limited English Proficient Compliance Review, EPA, Office of Civil Rights (December 15, 2011), http://www.epa.gov/civilrights/TitleVIcases/index.html#ldaf.

43. *Investigative Report for Title VI Administrative Complaint 0R9–15–95,* EPA, Office of Civil Rights (August 30, 2012).

44. Letter from environmental justice advocates, *supra* note 16, at 3.

45. Letter from environmental justice advocates to EPA administrator Lisa Jackson (July 3, 2012) (on file with author).

46. *Investigative Report for Title VI Administrative Complaint File No. 5R-98-R5,* EPA, Office of Civil Rights (1998) (*Select Steel* decision), http://www.epa.gov/ocr/docs/ssdec_ir.pdf.

47. *Draft Title VI Guidance for EPA Assistance Recipients Administering Environmental Permitting Programs (Draft Recipient Guidance) and Draft Revised Guidance for Investigating Title VI Administrative Complaints Challenging Permits (Draft Revised Investigation Guidance),* 65 Fed. Reg. 39,650 (June 27, 2000) (Draft Guidance Documents). The Draft Revised Investigation Guidance first articulated the presumption. *See* 65 Fed. Reg. at 39,680. The agency also noted that the presumption could be overcome, a curious statement since it had also noted that a Title VI investigation was not an adversarial proceeding that would directly involve participation by the impacted community.

48. See, for example, Richie (2013), describing challenges to greenhouse gas regulations.

References

Broder, John M. 2011. *Obama Abandons A Stricter Limit on Air Pollution*, New York Times, September 3, 2011, at A1.

Deloitte Consulting. 2011. Evaluation of the EPA Office of Civil Rights 2 (2011), http://www.epa.gov/epahome/pdf/epa-ocr_20110321_finalreport.pdf.

Environmental Law Institute. 2001. *Opportunities for Advancing Environmental Justice: An Analysis of U.S. EPA Statutory Authorities.* Washington, DC.

Environmental Protection Agency (EPA). 1990. *Draft New Source Review Workshop Manual.* Washington DC. http://www.epa.gov/ttn/nsr/gen/wkshpman.pdf.

Environmental Protection Agency (EPA). 2011. *Plan EJ 2014.* Washington DC., http://www.epa.gov/compliance/ej/plan-ej/index.html.

Environmental Protection Agency (EPA). 2012. Plan EJ Supplement: Advancing Environmental Justice Through Title VI, Draft. Washington, DC, http://www.epa.gov/compliance/ej/resources/policy/plan-ej-2014/plan-ej-civil-rights.pdf.

Environmental Protection Agency (EPA). 2013a. *Agency, Environmental Appeals Board Practice Manual.* Washington DC. http://yosemite.epa.gov/oa/EAB_Web_Docket.nsf/8f612ee7fc725edd852570760071cb8e/388bd7f5b1b242b385257bc-5004002b7/$FILE/Practice%20Manual%20August%202013.pdf.

Environmental Protection Agency (EPA). 2013b. Plan EJ Progress Report. Washington DC. http://www.epa.gov/environmentaljustice/resources/policy/plan-ej-2014/plan-ej-progress-report-2013.pdf.

Environmental Protection Agency (EPA). 2013c. Draft Technical Guidance for Assessing Environmental Justice in Regulatory Analysis (2013), Docket ID No. EPA-HQ-OA-2013–0320–0001, http://yosemite.epa.gov/sab/sabproduct.nsf/0/0f7d1a0d7d15001b8525783000673ac3!OpenDocument&TableRow=2.0.

Environmental Protection Agency (EPA). 2013d. Title VI of the Civil Rights Act of 1964: Adversity and Compliance with Environmental and Health-Based Thresholds. Washington, DC. http://www.epa.gov/civilrights/docs/pdf/t6.adversity_paper1.24.13.pdf.

Environmental Protection Agency (EPA). 2013e. Title VI of the Civil Rights Act of 1964: Role of Complainants and Recipients in the Title VI Complaints and Resolution Process. Washington, DC. http://www.epa.gov/civilrights/docs/pdf/complainants_role_issue_paper.pdf.

Environmental Protection Agency (EPA). 2014. *About the Environmental Appeals Board.* http://yosemite.epa.gov/oa/EAB_Web_Docket.nsf/General+Information/About+the+Environmental+Appeals+Board?OpenDocument (last visited Mar. 3, 2014). Environmental Protection Agency. No Date. Response to Public Comments on Plan EJ Strategy and Implementation Plans. Washington, DC. http://www.epa.gov/compliance/ej/resources/policy/plan-ej-2014/plan-ej-2011-comments-responses.pdf.

Environmental Protection Agency (EPA). No Date. Tribal Consultation and Coordinating Plan. Washington, DC. http://www.epa.gov/compliance/ej/resources/policy/plan-ej-2014/plan-ej-tribal-consult.pdf.

Environmental Protection Agency (EPA), Civil Rights Executive Committee. 2012. Developing a Model Civil Rights Program for the Environmental Protection Agency: Final Report. Washington, DC. http://www.epa.gov/epahome/pdf/executive_committee_final_report.pdf.

Foster, Sheila R. 2000. "Meeting the Environmental Justice Challenge: Evolving Norms in Environmental Decisionmaking," *Environmental Law Reporter* 30: 10922

Gauna, Eileen. 1996. Major Sources of Criteria Pollutants in Nonattainment Areas: Balancing the Goals of Clean Air, Environmental Justice, and Industrial Development. *Hastings West-Northwest Journal of Environmental Law and Policy* 3:379.

Gauna, Eileen. 2001. EPA at Thirty: Fairness in Environmental Protection. *Environmental Law Reporter* 31:10,528.

Hazardous Waste Consultant. 2013. Procedures for Appealing Permit Decisions before the Environmental Appeals Board. *Hazardous Waste Consultant* 31(2): 2.10.

Lazarus, Richard J., and Stephanie Tai. 1999. Integrating Environmental Justice Into EPA Permitting Authority. *Ecology Law Quarterly* 26:617.

LoPresti, Tony. 2013. Realizing the Promise of Environmental Civil Rights: The Renewed Effort To Enforce Title VI of the Civil Rights Act of 1964. *Administrative Law Review* 65:757.

McArdle, John. 2012. "EPA: Agency Creates Mapping Tool Aimed at Identifying 'Environmental Justice' Hot Spots," *Greenwire*, July 27, http://www.eenews.net/greenwire/stories/1059967993.

McCutchen, Gary, and Colin Campbell. 2011. "Horsefeathers!": Landmark Court Decision Directs EPA to Address Grandfathering, CAA One-Year Permit Processing Mandate. *Air Pollution Consultant* 21(4): 5.1.

National Academy of Public Administration. 2001. *Environmental Justice in EPA Permitting: Reducing Pollution in High-Risk Communities is Integral to the Agency's Mission.* Washington, DC.

National Academy of Public Administration. 2002. *Models for Change: Efforts by Four States to Address Environmental Justice.* Washington, DC.

National Academy of Public Administration. 2003. *A Breath of Fresh Air: Reviving the New Source Review Program.* Washington, DC.

National Environmental Justice Advisory Council. 2010. Nationally Consistent Environmental Justice Screening Approaches. May, pp. 7, 10, 13–14.

Richie, Alex. 2013. Scattered and Dissonant: The Clean Air Act, Greenhouse Gases, and Implications for the Oil and Gas Industry. *Environmental Law (Northwestern School of Law)* 43:461.

Shelby, Brendan C. 2013. Internal Agency Review, Authoritativeness, and Mead. *Harvard Environmental Law Review* 37:539.

Steinzor, Rena. 1996. Regulatory Reinvention and Project XL: Does the Emperor Have Any Clothes? *Environmental Law Reporter* 26:10527.

4

Assessing the EPA's Experience with Equity in Standard Setting

Douglas S. Noonan

When the Environmental Protection Agency (EPA) sets standards or takes other actions in promulgating rules to protect the environment, a host of equity concerns arise in various ways.[1] There are direct and indirect concerns about distributive and procedural fairness associated with the regulatory action. The standards set to curtail pollution follow from statutory authority granted to the EPA from congressional legislation and receive guidance from executive orders and other federal policy directives. The dual objectives of cleaning the environment and promoting a just distribution of environmental burdens pose challenges for an agency like the EPA in carrying out its mandates.

Although the EPA enjoys discretion in how it addresses environmental justice concerns, this discretion is constrained by both policy commitments and legal constraints. The core question addressed in this chapter pertains to how the EPA has used this discretion in standard setting. In particular, President Bill Clinton's Executive Order 12898 on Environmental Justice (EO 12898) called for the consideration of equity issues in standard setting. How has the EPA performed? Are environmental justice concerns now routinely considered when the agency sets pollution control standards? This chapter provides some answers. First, it outlines the nature of these concerns and how they arise, in theory and in practice, in the EPA's regulatory actions. Next, it reviews some of the available evidence on the distributive impacts of EPA regulations. Finally, the chapter concludes by offering some recommendations for federal environmental justice policy.

Policy Instrument Choice

Some background on the EPA's standard-setting role can help illuminate the opportunities and limits for environmental justice in environmental policy more broadly. When Congress passes an environmental law (e.g., the Clean Air Act [CAA], the Clean Water Act [CWA], the Resource Conservation and Recovery Act [RCRA], the Safe Drinking Water Act [SDWA], and the Toxic Substances Control Act [TSCA]), it typically delegates some discretion to the EPA to implement its mandate. This entails rulemaking by the EPA, and the rules it writes, programs it creates, and standards it sets are thus a vital part of pollution control efforts. These actions fill in the details of federal environmental policy that was impractical or undesirable for Congress to specify (and perhaps the EPA is more responsive). Yet the resulting regulation must derive from the statutory authority granted to the EPA and adhere to guidelines in the statute. The EPA simply lacks the authority to freely pursue its own goals (or the president's), and it cannot consider factors in its rulemaking if the statute directs the agency otherwise. Political concerns, such as White House edicts like EO 12898, and technological or institutional feasibility can play prominent roles in these scenarios. The EPA's rulemaking process is long and complex. It typically involves considerable development of alternatives, analysis of the implications of those options, collection of scientific information and public comment, review by multiple offices in the agency and administration, and publication in the Federal Register.

Historically, the EPA has typically implemented environmental laws by setting standards for allowable emissions, permissible technologies, or ambient pollution concentrations consistent with the statutes. Under the CAA, for instance, Congress directed the EPA to establish the National Ambient Air Quality Standards (NAAQS) according to some principles (e.g., protecting human health) while ruling out other criteria (e.g., economic cost), but it did not dictate the standards themselves. Similarly, Congress under the CWA established a permit system and dictated technology-based standards, but it left the EPA with the responsibility of specifying the limits and promulgating water quality standards. In sum, there is constrained discretion for the EPA in rulemaking.

The statutory constraints and political realities faced by the EPA help explain its past approach to environmental justice in standard setting.

The EPA's strategy for incorporating environmental justice and EO 12898 into its operations can be seen in the goals statement in its 1995 strategy document (EPA 1995) and in subsequent policy documents (e.g., EPA 2004). It has aimed to make sure no subpopulations suffered disproportionately (in terms of health or environment) from EPA actions and that everyone enjoys clean environments. Arguably, EPA policy has eschewed singling out justice for poor and minority populations but instead seeks "justice for all" (EPA 2004). This follows from the silence of major environmental statutes on justice and equity goals. From the EPA's perspective, the agency's environmental justice mission is not confined to acting only when disproportionate impact exists (EPA 2010). Their environmental justice goals are served by reducing pollution for disadvantaged groups, or for anyone in fact, regardless of disproportionate impacts.[2] This flexibility is coupled with an agency view that environmental justice policies ultimately only demand a consideration rather than ensuring any policy outcomes: "E.O. 12898 and the Agency's environmental justice policies do not mandate particular outcomes for an action, but they demand that decisions involving the action be informed by a consideration of EJ issues." (EPA 2010, p. 5) With a mandate to consider environmental justice and little authority to directly address it, the EPA affects distributional equity and procedural fairness in large part indirectly, through the broader policy approaches taken.

In carrying out federal environmental policy, there are several policy instruments generally available. Whether it is Congress selecting the instrument or delegating that decision to an agency like the EPA, the instrument choice involves designing a program to implement the environmental policy. And these program design decisions typically carry with them significant implications for environmental justice (see, e.g., Parry 2004). It is important to emphasize that the program design itself is a fundamental component of an environmental policy, not something taken as a given, and thus the "justice" of a federal action or inaction should be assessed in light of the alternative instruments available. Environmental policy instruments typically fall into several major categories—the broadest distinction being between conventional regulatory approaches and market-based approaches (Goulder and Parry 2008). The former category is often known as *command-and-control* regulation and traditionally has dominated federal environmental policy. The latter category emphasizes economic incentives and property rights. A third category, information-based policies, might usefully be introduced in this chapter as well.

Regulatory Approaches

Command-and-control approaches typically rely on prescribing either specific types of technology or performance standards for particular classes of pollution sources. These approaches typically revolve around standards and permits. The standards often dictate minimum performance of sources or the maximum ambient pollution concentrations for an area. Examples of performance standards include Best Available Control Technology (BACT) standards for air pollution and minimum technology requirements for waste disposal under RCRA. Examples of ambient concentration standards include the NAAQS for air quality and Total Maximum Daily Loads (TMDLs) for water quality. Standards-based approaches often have facially "just" properties, as they typically include uniform prescriptions across space and classes of individuals. Under the surface, however, a single standard that applies equally to all might have distributionally nonuniform impacts. Permits are issued to limit the sources. The CWA's National Pollutant Discharge Elimination System (NPDES) is a prominent permit system. Insofar as permits are conditional on sources meeting a technology or performance standard, the distinction between permits and standards as policy instruments blurs. Insofar as permits are tradable rights, then this regulatory approach constructs a market and shifts into the next category.

Market-Based Approaches

Economic incentive-based approaches rely on pollution taxes or trading systems. Establishing property rights can address pollution spillovers that arise when markets are incomplete, although this typically involves a regulator setting a "cap" for pollution or resource use before agents begin trading their shares of the quota. Trading under the lead phase-out program and under acid rain provisions of the CAA are prominent examples. Pollution taxes have similar properties in terms of economic efficiency. They also rely on a policymaker setting a "price" for pollution (rather than a quantity in the cap-and-trade system). These are rare in the United States at the federal level. Whether the pollution tax or the cap-and-trade approach is taken, the program design includes critical distributional questions of entitlements, how tax revenues get spent, and how the geography of pollution shifts as the economy transitions to cost-effective abatement.

Information-Based Approaches

Information-based approaches are relatively novel and experimental in environmental policy, relying on "nudging" or fostering pro-environmental behavior through more passive mechanisms like information provision, technical assistance and more overt marketing and education. The Toxics Release Inventory (TRI) and the EPA's Energy Star labeling program are two high-profile examples. Attempts to shame polluters or praise green stewards are included here, as well as agency-provided institutions or forums to facilitate voluntary action toward an environmental goal. These policies intersect with environmental justice concerns in several ways, partly because they have been deliberately used to counteract injustice and to shame (otherwise legal) activities and partly because information itself can be seen as not being distributionally neutral.

Environmental Justice Concerns in Standard Setting

The EPA's own guidelines distinguish between two primary environmental justice concerns: the distributional fairness of impact and the equity of meaningful access to the standard-setting process (EPA 2011c). This chapter includes procedural matters as a secondary concern because it focuses on outcomes or impacts rather than procedures. (An important component of procedural fairness is opportunity for public participation in decision making, which is the subject of chapter 6.) Regulatory actions raise environmental justice concerns over distributional equity of impacts through several channels. Table 4.1 outlines how environmental justice concerns manifest in agency standard setting and program design in many different ways, as discussed in more detail next.

Primary Environmental Justice Concerns

Traditionally, environmental justice covered the spatial distribution of pollution. At a basic level, pollution is a physical substance in space and time. The geography of pollution poses environmental justice concerns

Table 4.1

Environmental Justice–Related Issues and Challenges for EPA Rulemaking

Primary Concerns	
Spatial distribution issues	Hot spots from nature
	Hot spots from laissez-faire policies
	Hot spots from command-and-control policies
	Hot spots from market-based policies
Nonspatial inequities	Environmental harms contained in products, jobs
Equity in standards	Calibrating across subpopulations different susceptibilities and exposures
Compliance burden	Regressivity of policy tools
Secondary Concerns	
Information provision	Overcoming informational deficits, informing rulemaking
Equity economics	Grandfathering, windfall gains, and impacts on related markets
Equity in participation	Procedural equity affecting outcome equity
Equity and enforceability	Rule design affects equity of enforcement
Technicalities and definitions	Defining the scope (cross-media and cumulative exposures), relevant subpopulations, evidence

whenever that spatial distribution of environmental risks is not independent of the distribution of environmental justice–relevant subpopulations. The spatial correlation between environmental justice communities and environmental harms or gains—the EPA also pays attention to the beneficial impacts of its actions (EPA 2010)—becomes the primary concern of environmental justice in agency rulemaking. If pollution were spatially uniform, environmental justice concerns would be difficult to observe here. Yet pollution is rarely uniformly mixed in the environment. Given that pollution hot spots do arise, are they drawn to some types of communities more than others? (To simplify the discussion, I use the term *hot spot* to refer to any local peak in pollution levels, even if that peak is not dramatic.)

In previous work, I have emphasized the important distinction between inequities in existing environmental quality and inequitable impacts from an agency action (Noonan 2008). Descriptions of the spatial correlation between pollution and special classes of residents receive much attention (see chapter 1 for a review of this literature). Inequitable impacts, however, are far harder to identify because robustly attributing causality to an agency action in complex real-world settings requires a credible counterfactual. Assessing the distributional impact of an agency action typically requires understanding both how pollution and people are spatially distributed after promulgation of the new rule and how pollution and people would be spatially distributed in the absence of the new rule. The difference is the impact of the agency action.

In practice, robustly estimating this (causal) impact is difficult and not always sufficient. First, analytical resources (including data, expertise, and time) are often inadequate for the assessment task. Further, in a prospective study (where both the anticipated outcome and the counterfactual are speculative or hypothetical), the imprecision in the estimates may render findings inconclusive. Second, the EPA is also interested in addressing areas where the existing distribution is inequitable (EPA 2010). From this perspective, federal environmental policy faces environmental justice concerns whenever there is an inequitable distribution of hot spots regardless of its source.

Hot spots may arise naturally, from laissez-faire policy, from command-and-control regulation, or from market-based policies. First, consider hot spots from nature, which occur when inequities in the distribution of environmental quality is in the "background" or natural level of pollution that arises even without anthropogenic influences. Background levels of ozone, for instance, differ substantially around the United States (EPA 2006b). More than just environmental risks (e.g., severe weather, radon, mercury, arsenic, and disease vectors) with naturally varying background levels, environmental amenities generally vary spatially as well. From aesthetics to ecosystem services, environmental benefits are not naturally distributed uniformly. Nature itself creates hot spots.

Related to naturally occurring hot spots are those that arise in the absence of any deliberate environmental policy. If natural hot spots exist in the absence of humans, then laissez-faire hot spots are those that exist in the absence of any regulation. Pollution generated by economies that lack environmental policies may tend to cluster spatially. Concerns that market forces can lead to considerable clustering of polluting activities have long been linked to environmental justice concerns—especially when market forces that promote (or at least allow for) pollution hot spots coexist with incentives for subpopulations to also cluster into those areas (Hamilton 1993). This has been termed *environmental gentrification*, and it has been observed when lower income—correlated with ethnicity—limits the affordability of cleaner and greener neighborhoods, workplaces, and products. If low-cost host locations spatially cluster, such as along transportation corridors or where land or labor is cheap, so might we expect laissez-faire hot spots to develop. (Furthermore, if there is unregulated discrimination against subpopulations, and those subpopulations spatially cluster, then this might yield laissez-faire hot spots.) Recent work adds richer empirical details behind the mixed evidence of demographic sorting around nuisances (e.g., Banzhaf and Walsh 2010; Depro et al. 2012).

Even in the absence of natural or laissez-faire hot spots, regulation itself might create hot spots. Separate-use zoning as a way of reducing pollution exposure, for instance, attempted to create pollution hot spots by design. Regulatory tools like permitting or other standards-based approaches might also foster hot spots depending on their design and implementation. They might also remove preexisting hot spots. The NAAQS are designed to reduce (typically urban) air pollution hot spots and have the indirect consequence of shifting polluting industries to where it is cleaner (Greenstone 2002). There are multimedia tradeoffs as well, as regulatory attempts to reign in one sort of pollution can exacerbate other sorts of pollution (Greenstone 2003).[3]

Market-based approaches may also affect pollution hot spots. Pollution taxes, for instance, will have less impact in areas where pollution abatement costs are highest. Similarly, cap-and-trade systems may tend to see permits traded such that they actually create hot spots, especially if abatement costs tend to cluster geographically (Ringquist 2011; Turaga et al. 2011; Fowlie et al. 2012). This concern motivated the lawsuit brought against the California Air Resources Board in 2013 for its cap-and-trade approach to reducing greenhouse gas emissions. These market-based programs can be designed to account for or essentially penalize spatial clustering in those who continue to pollute.[4]

Even in the absence of spatial clustering of the pollutants, major environmental justice concerns may also arise due to the market distribution of risks where setting standards for pollution and environmental risks may have disproportionate impacts on subpopulations. If the pollution or risks are embodied in a product or in a production process, and if users of those products or workers in those production facilities tend to share similar demographic characteristics, such as income or race, then an environmental justice concern arises. The environmental justice community is not defined so much by commonality in residential location. If disadvantaged populations shop for similar items or work in similar facilities, they may be disproportionately affected. Examples like lead-based paint in homes and mercury contamination of fish involve fairly widespread product contamination, and also consumption patterns and exposures that correlate with income and ethnicity. Another prominent example is pesticide exposure among agricultural workers, who tend to share key traits like income, race, language, or nationality.

Primary environmental justice concerns in the standard-setting context also directly arise in the uniformity or variability of standards across subpopulations. Subpopulation-specific sensitivities and standards can complicate matters. Where the co-location of environmental risks and subpopulations directly informs the presence of injustice, the sensitivity of subpopulations to those risks need defining in order to assess the severity of the impact or exposure. Often, sensitivities are implicitly assumed to be the same across subpopulations. But this is not always the case, as air quality standards have been driven by the sensitivity of more vulnerable populations. While asthmatics are not typically a class associated with environmental justice, children and the elderly sometimes are. For air toxics, the CAA centers regulation on a threshold concept for the maximally exposed individual's risk, which accounts not at all for total population exposure or different subgroup susceptibilities (Goldstein 1995). Basing toxics regulation on maximum individual risk (MIR) appears facially neutral (inasmuch as all individuals enjoy the same MIR threshold), but MIR cannot account for different aggregate subpopulation risks.

Subpopulation-specific susceptibilities to airborne toxics are not available, except for children (Turaga et al. 2011). Discussions—with some controversy—about varying thresholds or standards based on ethnicity, ages, or even income continue today.

Finally, compliance burden and the costs of regulation may disproportionately affect certain groups (Bento 2013). Yet EO 12898 ignores this type of disproportionate impact. The EPA's own guidance documents for incorporating environmental justice concerns into an agency action do not acknowledge equity in compliance burden as an environmental justice concern (EPA 2010). Regardless, the economic impact of regulations can have important distributional inequities. Much of the public debates over major environmental regulations hinge on the compliance burden or potential regressivity of such a policy. Disadvantaged groups seeking relief (under the environmental justice banner) from disproportionate pollution burdens arising from economic growth would likely use the same banner to advocate for relief from disproportionate burden arising from efforts to clean up that pollution. These concerns might alter the regulatory actions the EPA favors (Nweke 2011).

Secondary Environmental Justice Concerns

There is also a set of secondary concerns that arise in the context of standard setting that are important to recognize, related to equity in information-based policies, economics, participation, enforcement, and definitions of scope. Each is briefly discussed next.

Information Provision Information-based policies have some ambiguity regarding their *impacts* on environmental justice communities. Reporting and dissemination programs, such as the TRI, are often touted as environmental justice–related policies. They can help redistribute environmental risks by empowering grassroots action (Fung and O'Rourke 2000). The EPA can make and alter rules about what information must get reported—such as expanding the chemical list as the agency did in 1994 and the industry sectors required to report as it did in 1997 (EPA 2014). This can better inform communities living near, say, mining operations or power plants by targeting them for reporting. In theory, more information might be good or bad for disadvantaged communities, depending on private actors' response to the new information. If information deficits are partly to blame for disparities, and the EPA's information programs can remedy them, then favorable impacts for environmental justice should be expected. But this is no guarantee. The TRI, for instance, may not help reduce toxic releases in poor or minority communities (Shapiro 2005).

Equity in Economics How agencies establish standards for pollution control has important economic implications that may have distributional consequences. In setting standards, one of the most controversial and high-profile issues involves grandfathering old technologies. For example, the older, dirtier, coal-fired power plants regulated under the CAA escape the more stringent New Source Performance Standards. Exempting older technologies typically undermines the environmental goals of the regulation, and it might have equity implications if those grandfathered technologies tend to pollute in disadvantaged communities or be owned by certain subpopulations. Similarly, issuing permits to pollute or use natural resources (e.g., fish and water) confers a valuable property right or asset to the recipient, and (unless they are sold by the government) entitlement recipients will enjoy windfall gains. Windfall gains from regulations and program designs that favor incumbent firms, technologies, or polluters likely have uneven distributional consequences in terms of which households enjoy those gains.

Standard setting may affect interdependent markets, which might raise environmental justice concerns. Air pollution regulations associated with nonattainment may shift manufacturing or construction jobs elsewhere, or it might disrupt industries like professional landscaping in ways that adversely affect some groups with disproportionate employment in that sector. Blocking the expansion of fossil fuel–based power generation might foster development of wind or solar farms that bring (potentially) negative consequences for host communities that might be low-income or minority.

Equity in Participation While the EPA considers equity in participation as a first-order concern (see chapter 6), it is mentioned here to emphasize one indirect consequence of equitable involvement in the standard-setting process. Meaningful participation by disadvantaged and affected communities in the standard-setting process might actually alter the rule itself (relative to what it would have been without such participation). Altering how standards are devised may itself influence the proportionality of their impact. For instance, the Community Advisory Groups (CAGs) and Technical Assistance Grants (TAGs) promoted under Superfund as Congress instructed the EPA to engage local communities more have led to tighter cleanup standards (Daley 2007). CAGs and TAGs were more likely to form in poor or minority communities, although minority communities tended to select less health-protective remedies. Meaningful participation may affect more than just procedural equity, and its influence on impact equity should not be presumed.

Equity in Enforcement The EPA's own guidance documents (e.g., EPA 2010) rightly remind rulemakers that designing regulations, setting standards, and requiring permits does not occur in a vacuum. Rulemakers are advised to think "downstream" in the policy process and to consider implementation issues, particularly monitoring and enforcement issues, and how they might invite or ameliorate environmental justice concerns (discussed in depth in chapter 7). Even at the standard-setting stage, potential inequities in enforcement—or enforcement realities that might lead to injustice—should be anticipated and the rule adapted to account for that likelihood.

Defining the Scope There are multiple operating definitions of what injustice would look like. Any attempt to measure or address environmental justice must clearly settle this matter, and agency rulemakers often have some discretion in how environmental justice is characterized. More important, how the standard is crafted can have important implicit consequences for environmental justice discussions and policy in that context. The scope of the regulation (which firms, regions, chemicals, etc.) sets boundaries that can themselves indirectly create hot spots or otherwise disproportionately affect impact. For example, TRI rulemakers decided numerous seemingly minor details—including reporting requirements, changing reporting rules over time, and exemptions. Reporting standards can have subtle implications for the data generated by this program— arguably one of the most important tools for promoting environmental justice.

Moreover, most standards are set for single pollutants or chemicals rather that the complex and often-interdependent array of chemicals.

This raises indirect yet arguably vital environmental justice concerns as disadvantaged communities are seen to suffer from multiple common afflictions, not just one environmental harm in isolation. Dealing with cumulative and cross-media effects in regulation involves a level of complexity and sophistication that conventional command-and-control tools have not yet embraced (nor have market-based approaches, for that matter). This is a major challenge for incorporating environmental justice into EPA rulemaking, in part because EPA regulations often rely on standards that set thresholds, thus limiting regulatory action against individual pollutants that fall below an acceptable threshold but may still pose a serious cumulative risk to a community affected by other pollutants (Nweke 2011).

Defining relevant subpopulations in the standard might also have justice implications. The EPA commonly interprets environmental justice as protective of *all* subpopulations, including nonminority or wealthy groups, and not just disadvantaged communities (Nweke 2011; EPA 2004). The relevant subpopulations enumerated in the Civil Rights Act are not the same as in EO 12898, and future regulations would need to specify which groups matter.[5] Exemptions for groups based on age, standards based on subgroups with particular medical conditions, or requiring notifications in multiple languages are all techniques that regulators have written into rules that define relevant groups. Generally, subpopulations are not explicitly singled out in environmental regulations, as their statutory bases do not generally require explicit treatment of different subpopulations.

If federal environmental policy starts tackling environmental justice concerns more deliberately, then how legislators or rulemakers choose to define the subpopulations, boundaries, and scales of measurement for environmental justice will have great consequences. As I argue elsewhere (e.g., Noonan 2008), these kinds of definitional issues can be pivotal in what constitutes evidence of injustice, and how we interpret that evidence. What constitutes a pattern of disproportionate impact remains a matter of enough concern that the Council on Environmental Quality (1997) has promulgated some guidance on it for National Environmental Policy Act (NEPA) reviews. Yet the guidance is problematic on its own. The technical questions of how to define environmental justice and how to measure disproportionate impacts (and, crucially, impacts of *what?*) will play a primary role once federal policy shifts from "considering" environmental justice concerns to requiring enforceable actions or outcomes related to environmental justice.

Environmental Justice in EPA Rulemaking in Practice

The EPA's rulemaking processes afford limited opportunities for incorporating environmental justice concerns. The use of such processes to constrain agency discretion (in particular the EPA's) are well documented (e.g., McCubbins et al. 1989). The elaborate set of procedures governing EPA rulemaking (EPA 2011b) include many key steps at which environmental justice concerns can be considered (EPA 2010, especially Appendix B), such as initiating the action (steps 1 and 2), devising preliminary and detailed analytic blueprints (steps 3 and 5), receiving guidance and approval from managers (steps 4 and 6), implementing the analytic blueprint (step 7), selecting the options (step 8), preparing the rule and its documentation of efforts to consider environmental justice (steps 9 and 10), soliciting public comment (step 15), and developing the final rule (step 16). This goes far beyond public participation and the "notice and comment period" within the rulemaking process. Such comment periods are not known to give voice to environmental justice concerns (Golden 1998; Coglianese 2006).

How the EPA addresses environmental justice in its rulemaking can be thought of in several other respects. There are formal rules, institutions, and guidelines that explicitly cover environmental justice in the agency. EPA actions (as well as its inactions) reveal yet more about how the agency addresses environmental justice in its rulemaking.

Organizational Structure Implementing environmental justice in the EPA is not a straightforward practice from an organizational standpoint. The first place that one might look for EPA's activities regarding environmental justice would be its Office of Environmental Justice (OEJ). The OEJ coordinates the EPA's environmental justice–related efforts, but it does not promulgate rules and standards. Other rule-making offices at the EPA do this as part of programs.[6] Other offices promote environmental justice in the development of science tools (Office of Research and Development) or information tools (e.g., Office of Policy and Office of Environmental Information), or in fostering compliance (elsewhere in the

Office of Enforcement and Compliance Assurance). The Office of Civil Rights (OCR) resides in the Office of the Administrator to handle Civil Rights Act of 1964 compliance. Offices of General Counsel and of Inspector General exist as separate offices in the EPA as well. Existing programs in various EPA offices already pursue environmental justice goals (broadly defined), such as the Office of Solid Waste and Emergency Response's Community Engagement Initiative, the Office of Water's Urban Waters program, the Office of Enforcement and Compliance Assurance's National Enforcement Initiatives, the Office of Air and Radiation's Air Toxics Rules, and the Office of International and Tribal Affairs's U.S.–Mexico Border Program. (EPA 2011c).

One of the most important organizational aspects of environmental justice policy in the EPA is the National Environmental Justice Advisory Council (NEJAC), established in 1993. NEJAC representatives, drawn from diverse sectors of society, meet a couple times per year to discuss environmental justice and advise the EPA administrator. NEJAC recommendations include topics like incorporating cumulative impacts into permitting, monitoring at schools, and better collaborating with communities, and its recommendations over the past 20 years address very relevant aspects of EPA rulemaking. However, while NEJAC is one of the higher-profile, institutionalized aspects of environmental justice at the EPA, it is ultimately an advisory committee.

Rulemaking Opportunities   NEJAC (2002) observed a decade ago that, except for NEPA, the EPA's statutory authority for taking actions based on environmental justice concerns is limited. The major federal environmental statutes do offer some latitude for consideration of equity and justice issues, however. Readers are directed to one of the several listings of opportunities for the EPA to exercise its discretion under current statutes to incorporate environmental justice considerations (e.g., NEJAC 2002; Lazarus and Tai 1999; Guzy 2000; EPA 2011d). As indicated in Table 4.2, the EPA has exercised some of this discretion. Table 4.2 lists select opportunities under statutes such as the CAA, the CERCLA, the RCRA, and the Federal Insecticide, Fungicide, and Rodenticide Act (FIFRA) where the EPA has incorporated environmental justice or some other special consideration for a subpopulation in its rules and standards. Grants, procurement, and review programs are omitted from this table.

Table 4.2

Select Environmental Justice Issues Present in Federal Statutes for EPA

CAA	Program Element	Environmental Justice–related Issue	Agency Opportunity or Action
CAA	Section 109	Public health	Flexibility in setting NAAQS, establishing the monitoring network
CAA	Sections 211, 213	Public health	Flexibility in controlling fuels and nonroad engine emissions
CAA	Sections 110, 165, 172, 173	Permit requirements	Deny permits based on disproportionate impact, possibly require environmental justice consideration in permitting
CERCLA	National Contingency Plan	Protectiveness of human health	Flexibility in choosing remedial alternatives
CWA	Section 304	Highly exposed populations	Flexibility in setting standards for subpopulations eating more fish
CWA	Section 402	Oversight of state permits	Flexibility in considering environmental justice in reviewing/objecting to state-issued permits
CWA	Section 404	Adverse discharge effects	Can object to or veto the permits from the U.S. Army Corps. of Engineers for environmental justice reasons
EPCRA	Section 313	Information reporting	Flexibility in what and how TRI info is reported; flexibility in adding chemicals or facilities
FIFRA	Section 3	Economic, social, and environmental costs and benefits	Flexibility in considering distribution of costs, benefits of registering pesticides
FIFRA	Worker protection standards, labeling	Affected populations	Sets rules for labeling and notification, possibly targeting specific groups
RCRA	Section 3005	Public health	Flexibility in denying permits for environmental justice reasons
SDWA	Underground injection control program	Public health	Flexibility in denying permits for environmental justice reasons
TSCA	Section 6	Social impact	Any regulatory action under TSCA could be based on EJ
TSCA	Section 305	Homes of low-income persons	Statute instructs EPA to target assistance based on income

Note: Examples where EPA has previously undertaken some action that relates to environmental justice concerns (explicitly or indirectly), from EPA (2011d).

Crucially, because legislation authorizes the EPA to base standards on protecting public health rather than on environmental justice, the EPA has successfully used the public health arguments to justify standards based on subpopulations in the past. For example, the EPA opted to define the primary standards for the NAAQS in part to protect sensitive subpopulations (asthmatics, children, elderly, etc.) as part of its mandate to set standards to protect public health. It set standards based in part on the different dose-response relationship for these sensitive subpopulations, but it has not established or acted on a different sensitivity by race or income. Likewise for its leaded gasoline phaseout, the differentially harmful effect of airborne lead for children justified the EPA's aggressive standards. Debate continues over using children's sensitivity to lead exposure to revise standards for dust and soil, where the costs of tightening standards currently blocks change. The EPA has not yet employed a similar tactic for populations based on common environmental justice categories, but such an approach *may* be permissible (EPA 2011d). Even when permissible, challenges with the feasibility of incorporating environmental justice concerns under existing statutes remain a problem (NEJAC 2002) or a matter of "legal risk" (EPA 2011d). If new rules were based on environmental justice–related criteria, then they may become stranded in federal court and stricter pollution controls for everyone get delayed.

Consider the 2008 RCRA Hazardous Waste Definition of Solid Waste Rule as an example of the EPA explicitly considering environmental justice concerns in rulemaking (EPA 2011a). Specifically, the EPA investigated disproportionate impact resulting from a rule clarifying how the solid waste definition applies to hazardous material intended for recycling. The EPA had some discretion over its hazardous waste rules under RCRA and opted to incorporate environmental justice concerns into its regulatory impact analysis (RIA). Its process solicited public discussions about the methodology before drafting its environmental justice analysis. Again, this is an example of the EPA *considering* environmental justice concerns through its rule-making process while the final regulatory option selected might not explicitly address those concerns.

Other instances of the EPA exercising discretion in incorporating environmental justice into its rulemaking are few and far between. For instance, under the TSCA, the EPA recently extended its lead-based paint rules to all pre-1978 child-occupied facilities and housing under the explicit premise of pursuing environmental justice (EPA 2011d). The TSCA also contains provisions that explicitly benefit low-income groups facing home radon and lead-based paint risks, and the EPA carries out those grant programs. The EPA has set standards under the CWA to protect the highly sensitive group of subsistence fishers, as they are an "existing use" that must be protected. Although somewhat indirect, water quality standards set for subsistence fishers can protect an environmental justice community of Native American populations with high subsistence fishing rates. EPA's pesticide-related labeling rules and tolerance levels also reflect a sensitivity to disadvantaged communities (Gerber 2002). Some underground injection permits have been denied and some section 404 (filling of wetlands) permits have been vetoed with reference to environmental justice, at least in some instances (EPA 2011d). These cases seem more like exceptions rather than the rule, though. Overall, EO 12898 is invoked proactively only rarely in federal rulemaking (Gerber 2002).

EPA's Track Record

Environmental justice advocates often argue that, whatever discretion they have, agencies like the EPA have substantially neglected their environmental justice missions in practice. In reviewing the EPA's track record for incorporating environmental justice concerns in its rulemaking, Rubin and Raucher (2010) highlight how it has not followed up on its tough talk on environmental justice. Only rarely has an environmental justice analysis been conducted prior to the EPA promulgating a rule [less than 7 percent of the time in the first seven years under EO 12898—and largely under the same Clinton administration that issued the order (Gerber 2002)]. The weak role of environmental justice in EPA rulemaking was not expected to fare any better under the Bush administration that followed. Accordingly, the EPA's own Office of Inspector General declared:

EPA has not fully implemented Executive Order 12898 nor consistently integrated environmental justice into its day-to-day operations. EPA has not identified minority and low-income, nor identified populations addressed in the Executive Order, and has neither defined nor developed criteria for determining disproportionately impacted. Moreover, in 2001, the Agency restated its commitment to environmental justice in a manner that does not emphasize minority and low-income populations, the intent of the Executive Order. (EPA 2004, p. i)

The next year, the Government Accountability Office (GAO) reviewed rulemaking under the CAA and concluded that "EPA generally devoted little attention to environmental justice" (GAO 2005, p. 3). The GAO criticized the EPA for a lack of guidance and lack of training of rulemakers with respect to environmental justice, not having necessary data to analyze environmental justice impacts, inconsistent and contradictory discussion of environmental justice, and not being sufficiently responsive or forthcoming about environmental justice concerns. A year later, the EPA's Office of Inspector General again reported:

Our survey results showed that EPA program and regional offices have not performed environmental justice reviews in accordance with Executive Order 12898. Respondents stated that EPA senior management has not sufficiently directed program and regional offices to conduct environment justice reviews. (EPA 2006a, p. 5)

Although there are thousands of scholarly articles about environmental justice, the vast majority of the empirical literature does not attribute evidence of environmental inequities to specific EPA regulatory actions.[7] Rather, scholars describe distributional equity in pollution more generally. A much smaller set of articles examines the narrower question of distributional impacts *of EPA actions*. A handful of these studies that find evidence on the link between EPA actions and environmental justice are summarized in Table 4.3. Notice how half of the studies in Table 4.3 address implementation of the Superfund program rather than the impacts of the EPA promulgating regulations.

The rarity of these studies reflects the difficulty in identifying causal impacts of EPA actions on different subpopulations and, perhaps, how distributionally just impacts are not a high priority for evaluators. Overall, the peer-reviewed academic literature has established relatively little about justice of the impacts of EPA rulemaking. For the Superfund, there is mixed evidence on the remedial decisions and pace, depending on which study is consulted. Earlier studies found no or minor correlations with income and race. Later studies suggest a role for income not altogether consistent with conventional environmental justice narratives. The idea that supporting and including community groups affect remedies presents an interesting example of meaningful participation also affecting treatment or outcomes. Fish advisories appear to have fairly universal impacts even when aimed at and based on subpopulation risks. This

Table 4.3

Select Evidence of the Unequal Impacts of Federal Actions

Authors		Research Question	Finding
Gianessi et al. (1979)	CAA	Do uniform CAA standards yield uniform results?	No. The poor appear to gain the most.
Hird (1990)	CERCLA	Is the cleanup pace or spending at National Priorities List (NPL) sites correlated with neighborhood income?	Neither are.
Hamilton (1993)	Hazardous waste processing facilities	Did the post-CERCLA regulatory regime change the siting of hazardous waste facilities?	No longer drawn to counties with more minorities; collective action explained more.
Gupta et al. (1996)	CERCLA	Do demographics affect EPA remedial decisions?	No. Permanent remedies were not favored differently in minority or poor areas.
Sigman (2001)	CERCLA	Do demographics affect listing, cleanup pace?	Somewhat. Community income affects pace; progress is faster in areas with more Hispanics.
O'Neil (2007)	CERCLA	Do neighborhood demographics predict the likelihood of a proposed site getting listed to the NPL? Did EO 12898 increase the equitability of the Superfund program?	Proposed sites in poor and minority tracts are less likely to be listed. Post-EO 12898, sites in minority tracts are even less likely to be listed.
Daley (2007)	CERCLA	Do EPA-supporting local citizen groups affect remedial decisions?	Yes. Forming CAGs and TAGs leads to more health-protective cleanup approaches.
Noonan (2008)	CERCLA	Does the racial composition or income of a neighborhood predict NPL deletions?	Deletions uncorrelated with race, less likely to occur in wealthier areas.
Shimshack and Ward (2010)	Mercury advisories in fish	Did advisories alter consumption? Differently for different groups?	Fish consumption fell, even for groups not at risk.
Baryshnikova (2010)	Air emissions at pulp and paper mills	Does regulatory pressure yield inequitable impacts on plant abatement?	Children and high school dropouts enjoy less abatement; no difference for minorities and poor.
Ringquist (2011)	CAA	Does the SO2-trading regime transfer pollution to minority communities?	No. Minority communities received fewer imports.

might be seen as having adverse impacts on nonminority populations. For air quality, it looks like, if anything, CAA rules have favored minorities. One of the more striking results is that of Ringquist (2011), who does not find evidence of injustice with respect to the hot spots generated by the SO2 trading program.

Some Progress

Despite the EPA's lackluster track record for making environmental justice a prominent part of its regulatory process, the agency has received praise for making environmental justice a priority during the presidency of Barack Obama (NEJAC 2011b). There have been a number of visible efforts to address environmental justice in program design and standard setting.

First, there has been some effort to vary *standards based on local circumstance*. By and large, standard setting is centralized at the EPA, where a single standard is promulgated nationwide and rules exist for implementation of the rule by others (states in particular). The Superfund program is one exception to this, where decision making is rather decentralized and cleanup standards are set largely on a site-by-site basis. In that decentralized arena, the EPA's standard-setting practices are easier for local community groups (in particular disadvantaged groups) to influence. Daley (2007) points to this being the case, as a sort of negotiation has occurred with local communities over cleanup standards.

Second, the EPA has exhibited some consideration of equity in air quality regulation. For example, under its authority to set NAAQS to protect public health, the agency opts to consider the health risks to sensitive populations, "which often provides an important opportunity to consider the health impacts on minority, low-income, and indigenous populations without an additional requirement that those impacts are disproportionate" (EPA 2010, p. 5). In addition, the EPA recently tightened the NAAQS for nitrogen dioxide (NO_2), and these new rules included new provisions for determining compliance. The new standards require 40 additional NO_2 monitors to be placed in susceptible and vulnerable communities to supplement the usual monitors (EPA 2010). This demonstrates how other aspects of rulemaking can be used to pursue environmental justice goals even without fundamentally altering a uniform standard approach. As Gina McCarthy, who replaced Lisa Jackson as EPA administrator in 2013, observes, it took 15 years after the EO for "a rule [to] actually recognize the issues of disproportionate impact" (NEJAC 2010, p. 92).

Third, the EPA has worked to improve its screening tools. If EPA actions have disproportionate impacts because of a lack of information within the agency, then better information might help it mitigate its own disparate impacts. The EPA has been drafting the Environmental Justice Strategic Enforcement Assessment Tool (EJSEAT) as an internal tool to help target enforcement efforts. New tools like GeoPlatform, promised as part of *Plan EJ 2014*, might be a major step by the the EPA to address criticisms that its own rulemaking processes are underinformed about their environmental justice consequences (EPA 2004, 2006a). EJView is another screening tool for external stakeholders.

Fourth, the primary strategy for implementing environmental justice in federal agencies remains one of seeking improved environment quality (for all, but perhaps especially for overburdened groups) rather than addressing disparate impacts (Binder et al. 2001). This was the explicit goal of the EPA's past administrators, and—despite criticisms that making environmental justice "for everyone" amounts to reinterpreting EO 12898 (EPA 2004)—remains embedded in *Plan EJ 2014*. The language used by the agency regarding environmental justice has recently shifted to emphasize targeting "overburdened communities," but little has changed in actual rulemaking.

Last, and perhaps most noteworthy, the EPA has addressed rulemaking as part of its *Plan EJ 2014* (EPA 2013). In the 20 years since EO 12898, the agency's performance in terms of gearing its mission around environmental justice must be rated poor. Facing major constraints— virtually no statutory mandates, courts circumscribing the scope of the Civil Rights Act, unfavorable political conditions, and lack of public support—the EPA spent the last 20 years picking its battles in ways that left environmental justice as a mere "consideration." Yet 2014 might mark a new era for environmental justice at the EPA. *Plan EJ 2014* lays out many goals and several milestone deadlines (some of which have already been missed). Ultimate implementation of *Plan EJ 2014* may be difficult to assess, given its "extremely general" nature and lack of metrics for accountability (NEJAC 2011b). In that regard, not much has changed for the EPA. Even the rhetoric has not evolved much, as evidenced by the fact that *Plan EJ 2014* omits the phrase "on minority populations and low-income populations" when quoting EO 12898's mandate that agencies address disproportionate impacts (EPA 2011b, p. 1).[8] On the other hand, *Plan EJ 2014* is not some overly ambitious and radical departure. Like many NEJAC recommendations, its tangible impact on EPA regulations is not yet established. Still, policy tools are being developed currently by the EPA, and progress at incorporating environmental justice into rulemaking is positive. More attention from rulemakers within the agency is being directed to addressing cumulative exposure in standard setting (EPA 2011b; EPA 2013).

How this will evolve in practice remains to be seen. In May 2013, the EPA released its *Draft Technical Guidance for Assessing Environmental Justice in Regulatory Analysis*. While still in draft form, this guidance document seeks to inform and improve internal EPA environmental justice analyses conducted in its various program offices as new rules are being considered. It is noteworthy in several respects. First, it marks the first effort for the EPA to formalize its internal approach to empirically analyzing environmental justice issues related to its rule-making activities. When and in what form this guidance document ultimately takes remain unknown, but important first steps have been made. Second, the draft draws on EO 12898 almost exclusively in framing its environmental justice analyses. As a result, it emphasizes the environmental justice groups identified in that order (i.e., minorities; low-income, indigenous populations; and subsistence fishers), and it directs the analytical focus on the equity implications of anticipated impacts of regulatory actions rather than more studies of legacy injustices. These formal analyses would merely describe the differential effects, if any, as matters of fact. It also explicitly links the EPA's own environmental justice analyses with its other RIAs (e.g., benefit-cost analyses). Linking internal technical guidance with the EPA's interim policy document (EPA 2010) on when to consider environmental justice in rulemaking may lead to more consistency and routine conduct of environmental justice analyses. Notwithstanding the details in the final guidance documents, the fact that the EPA has moved forward on this path toward formalizing and integrating environmental justice analyses into its rulemaking marks a major milestone in implementing EO 12898 at the agency.

More than just paying lip service to environmental justice and EO 12898, efforts to bring more consistency, transparency, and technical expertise to environmental justice considerations in EPA rulemaking may benefit environmental justice advocates. Maguire and Sheriff (2011) emphasize the inescapably normative element to incorporating environmental justice into EPA rulemaking in practice and how consensus has not been reached over how make best use of the available scientific evidence. Such consensus is not imminent, especially in the current U.S. political arena. Yet more and better risk assessment and environmental science has been seen to help environmental justice advocates' cause (Nweke et al. 2011, Kreger et al. 2011). Even when risk assessment methods are poorly suited to detecting disproportionate exposure for key subgroups, such as tribal populations and the geographically dispersed (Burger et al. 2010), often this is more about feasibility and data limitations rather than science actually not serving the interests of environmental justice advocates.

Conclusions

The EPA's mission to promote a cleaner environment, and the statutes that give it the specific authority to carry out that mission, largely get translated into a rather narrow, yet powerful notion of environmental justice. To the extent that clean environments are a human right and justice is served by improving environmental quality across the board, the actions taken by federal environmental regulators is directly connected to an environmental justice mission. For all the talk (and an executive order) addressing disproportionate impact and the idea that injustice involves differential burdens, the overwhelming bulk of the EPA's activities and efforts have targeted reducing pollution levels and *not* equalizing burdens. In a sense, the best thing that the agency has done for environmental justice has been to push for a cleaner environment for all. At least to date, it has done relatively little to spare particular subgroups from the burdens of regulations or of pollution, except insofar as standards may be set with an eye to some sensitive subpopulations.

This emphasis on overall pollution reduction rather than burden equalization follows directly from the statutory mandates given to the EPA by Congress. The aims of major federal environmental legislation have not been centered on promoting equity, but rather on reducing polluting behavior and reducing ambient pollution concentrations. If anything, the federalization of environmental policies has pushed more for uniformity in standards and policy applications—one-size-fits-all approaches—at the expense of more localized and variegated policies that might address burden inequities. (This is not by chance, as the freedom to particularize policies and standards is no guarantee that inequitable burdens will be ameliorated. If anything, the race-to-the-bottom concerns fueled expectations that inequitable burdens would be exacerbated by customized policies.) As statutes called for improved environmental quality, it was only a later EO 12898 that called for attention to disproportionate impacts. The teeth lacking in an executive order and the constrained scope of the Civil Rights Act put equity in a distinctly subsidiary position compared to overall pollution reduction directives in legislation. Seen in this light, any overt, nontrivial attention paid to environmental justice by the EPA in administering its air, water, and other pollution control activities might be considered impressive.[9]

Of course, the EPA enjoys (or suffers) considerable discretion in establishing the rules that implement the environmental statutes. This gives the EPA *some* latitude to incorporate environmental justice concerns in their activities. But the discretion can cut both ways, and where the political context is not favorable to environmental justice, we have seen EO 12898 lose its appeal (Gerber 2002). If Gina McCarthy's remarks in 2010 (NEJAC 2010) are any indication, there is little reason to expect the EPA to launch into creative and aggressive use of agency discretion in integrating environmental justice concerns into rulemaking, to do more than "consider" environmental justice, and to actively identify and eliminate disproportionate impacts on poor and minority communities. Instead, we might well expect more use of information-based approaches, avoidance of trading programs, and possibly localized targeting of sources for carbon dioxide (CO2) emissions. Furthermore, the flexibility that bends environmental justice to political winds can also make it a convenient tool to thwart other goals. (For instance, tighter pollution standards with inequitable distribution of benefits or regressive compliance costs might pose scenarios where environmental justice concerns can be used to block stronger environmental protections.) This explains some of the reluctance to require more than consideration of environmental justice in standard setting, lest it constrain the EPA's ability to improve overall environmental quality.

To go further than seeking a general improvement in environmental quality will require more creative use of its rule-making discretion, which is happening now in several dimensions. The EPA can accomplish this by requiring special monitoring. It can try to peg its standards to vulnerable subpopulations. In addition, it can write rules with flexible, market-based incentives and then regulate the markets to prevent hot-spot formation.

Going further still likely requires that it deal with cumulative exposure, deal with compliance costs, and adapt regulations (including market mechanisms) to account for hot spots. But a stronger and sustained public mandate for justice is likely a necessary condition for real progress on federal environmental justice policy, as environmental justice in rulemaking appears tied to political whims. Promulgating regulations to favor environmental justice goals can still be accomplished in the right political circumstances, especially with creative rulemaking. For instance, the perception that emissions trading yields hot spots may not manifest itself in practice (Ringquist 2011; Fowlie et al. 2012), and zonal trading programs like the Regional Clean Air Incentives Market (RECLAIM) in southern California can regulate pollution markets for environmental justice goals. Other federal policy efforts can empower relevant subpopulations to make better choices through informational programs and through endowing environmental justice communities with property rights to clearer environments (Banzhaf 2009).

The EPA should take the leadership in drafting scientifically grounded protocols and technical guidance for conducting environmental justice analyses. Two recent NEJAC recommendations are especially prudent and pressing:

• "Cumulative environmental impacts, permitted or not, must be addressed and mitigated within existing and new permits, regardless of permit type." (NEJAC 2011a, p. 2)

• "Evaluate the extent to which current practices and policies actually are contributing to poor environmental quality and health outcomes in certain communities." (NEJAC 2011b, p. 6)

Protecting public health has been interpreted to allow for setting standards based on sensitive populations. This puts pressure on integrating cumulative exposure models into standard setting, under the rationale that environmental justice communities suffer extra multimedia exposures and hence have greater sensitivities. A logic that went beyond physical susceptibility to a broader notion of vulnerability, which might include economic and social resources upon which self-protection and recovery depend, would further expand the opportunity to develop standards that protect environmental justice communities.

Recent progress at the EPA is highlighted by the report EPA's Action Development Process: Interim Guidance on Considering Environmental Justice During the Development of an Action (EPA 2010), the draft technical guidance document on conducting environmental justice analyses of EPA rules, and the initiation of the Science Advisory Board review process. That these efforts are being undertaken marks major progress for an agency that arguably evaded EO 12898 for many years. The process is slow and must still be graded as "incomplete" until those guidance documents become finalized. Of course, the ultimate impact on EPA operations and standard setting in practice will be limited by technical feasibility (i.e., performing high-quality, ex ante environmental justice analyses is a difficult task) and political concerns.

Certainly, federal action on environmental justice needs to take a broader view of what constitutes quality of life for affected and disadvantaged populations, not to mention society as a whole, and that means accounting for environmental, health, social, *and* economic impacts of agency actions (NEJAC 2011b). The economic consequences in environmental settings are typically crucial, not just because the economic costs of regulation are often quite real (in terms of higher prices, lost jobs, and other factors), but also because appreciating the economics is vital to appreciating how people and firms will react to regulatory actions.

Notes

1. When it comes to assessing the impact of EO 12898 in federal environmental policy, the EPA deservedly attracts the most attention. Some 95 percent of federal rules citing EO 12898 come from the EPA (Gerber 2002).

2. Framing environmental justice this way can be especially important in avoiding instances where the EPA needed to identify the distributional effects of its policies, they were shown to be regressive, and this regressivity limited the EPA's ability to move the regulations forward.

3. Muller et al. (2009) give an example of how reducing some NOx emissions would actually increase ozone and particulate matter pollution, suggesting that regulating NOx emissions in some areas can actually worsen the hot spots for some pollutants.

4. The recent federal court rulings have effectively stopped the EPA's SO2 trading program, ostensibly because it did not have state-specific caps and, by permitting interstate trades, was neglecting the relationship between sources and receptors. In a sense, it was the lack of mechanisms to prevent permits from concentrating in some states or near state borders that doomed the "grand experiment" in pollution trading (Schmalensee and Stavins 2013).

5. EO 12898 limits its attention to avoiding disproportionate impact on minority and low-income populations. The Civil Rights Act targets discrimination based on race, color, or national origin.

6. This includes the Clean Air Act (CAA), Clean Water Act (CWA), Resource Conservation and Recovery Act (RCRA), Comprehensive Environmental Response, Compensation, and Liability Act (CERCLA), Toxic Substances Control Act (TSCA), Safe Drinking Water Act (SDWA), and Emergency Planning and Community Right-to-Know Act (EPCRA).

7. While some of the literature might ascribe responsibility to EPA oversight or regulation, the research designs employed are rarely if ever capable of linking the equity of the outcome to specific EPA actions or policies (see Noonan 2008). These studies' data on pollution or polluters are related to EPA only out of convenience because EPA collects and makes the data available, rather than because they are analyses of EPA's actions per se.

8. Consider the remarks by McCarthy at the January 2010 NEJAC meeting on "new strategies" for improving health in environmental justice communities (NEJAC 2010). She noted that CO_2 emissions reporting requirements could be used as a tool for EJ, stronger NO_2 NAAQS, and more "justice for all" examples. Arguably, the only novel and EJ-specific strategy she listed was the new NO_2 monitors.

9. The reality of the EO version is that it inhibits EPA rulemaking by reducing the agency's latitude if it precludes regulating in ways that have disparate impacts. Embracing that part of environmental justice will limit the EPA's scope. Including equity in compliance costs under the banner of environmental justice may further restrict the EPA's scope.

References

Banzhaf, H. Spencer. 2009. And Environmental Justice for All But Liberty Comes First. *PERC Report* 27 (2): 27–30.

Banzhaf, H. Spencer, and Randall Walsh. 2010. Segregation and Tiebout Sorting: Investigating the Link between Investments in Public Goods and Neighborhood Tipping. NBER Working Paper No. 16057.

Baryshnikova, Nadezhda V. 2010. Pollution abatement and environmental equity: A dynamic study. *Journal of Urban Economics* 68:183–190.

Binder, Denis. 2001. A Survey of Federal Agency Response to President Clinton's Executive Order No. 12898 on Environmental Justice. *Environmental Law Reporter* 31:11133.

Bento, Antonio. 2013. Equity Impacts of Environmental Policy. *Annual Review of Resource Economics* 5:181–196.

Burger, Joanna, Stuart Harris, Barbara Harper, and Michael Gochfeld. 2010. Ecological Information Needs for Environmental Justice. *Risk Analysis* 30 (6): 893–905.

Coglianese, Cary. 2006. Citizen Participation in Rulemaking: Past, Present, and Future. *Duke Law Journal* 55 (5): 943–968.

Council on Environmental Quality. 1997. *Environmental Justice: Guidance under the National Environmental Policy Act.* Washington, DC: Executive Office of the President.

Daley, Dorothy. 2007. Citizen Groups and Scientific Decisionmaking: Does Public Participation Influence Environmental Outcomes? *Journal of Policy Analysis and Management* 26 (2): 349–368.

Depro, Brooks, Christopher Timmins, and Maggie O'Neil. 2012. Meeting Urban Housing Needs: Do People Really Come to the Nuisance? NBER Working Paper No. 18109.

Ellerman, A. Denny, Paul L. Joskow, and David Harrison, Jr. 2003. Emissions Trading in the U.S.: Experience, Lessons, and Considerations for Greenhouse Gases. Pew Center on Global Climate Change, Washington, DC. http://www.pewclimate.org/docUploads/emissions_trading.pdf.

Environmental Protection Agency (EPA). 1995. *Environmental Justice Strategy: Executive Order 12898.* Report No. EPA-200-R-95–002. Washington, DC: Environmental Protection Agency.

Environmental Protection Agency (EPA). 2004. *EPA Needs to Consistently Implement the Intent of the Executive Order on Environmental Justice.* Report No. 2004-P-00007. Washington, DC: Environmental Protection Agency.

Environmental Protection Agency (EPA). 2006a. *EPA Needs to Conduct Environmental Justice Reviews of Its Programs, Policies, and Activities.* Report No. 2006-P-00034. Washington, DC: Environmental Protection Agency.

Environmental Protection Agency (EPA). 2006b. *Air Quality Criteria for Ozone and Related Photochemical Oxidants.* vol. I., 3–48. Washington, DC: Environmental Protection Agency.

Environmental Protection Agency (EPA). 2008. *Environmental Justice Resource Guide: A Handbook for Communities and Decision-Makers.* Washington, DC: Environmental Protection Agency, Pacific Southwest/Region 9.

Environmental Protection Agency (EPA). 2010. *EPA's Action Development Process: Interim Guidance on Considering Environmental Justice During the Development of an Action.* Washington, DC: Environmental Protection Agency.

Environmental Protection Agency (EPA). 2011a. *Environmental Justice Analysis of Proposed Definition of Solid Waste Rule: Draft for Public Comment. Office of Solid Waste and Emergency Response/U.S. Environmental Protection Agency.* Washington, DC: Environmental Protection Agency.

Environmental Protection Agency (EPA). 2011b. *EPA's Action Development Process*. Washington, DC: Office of Policy, Environmental Protection Agency.

Environmental Protection Agency (EPA). 2011c. *Plan EJ 2014*. Washington, DC: Office of Environmental Justice,.Environmental Protection Agency.

Environmental Protection Agency (EPA). 2011d. *Plan EJ 2014: Legal Tools*. Washington, DC: Office of General Counsel, Environmental Protection Agency.

Environmental Protection Agency (EPA). 2013. *Plan EJ 2014 Progress Report. February 2013*. Washington, DC: Environmental Protection Agency.

Environmental Protection Agency (EPA). 2014. TRI Program Timeline. http://www2.epa.gov/toxics-release-inventory-tri-program/tri-program-timeline-0

Fowlie, M., S. P. Holland, and E. T. Mansur. 2012. What Do Emissions Markets Deliver and to Whom? Evidence from Southern California's NOx Trading Program. *American Economic Review* 102 (2): 965–993.

Fung, Archon, and Dara O'Rourke. 2000. Reinventing Environmental Regulation from the Grassroots Up: Explaining and Expanding the Success of the Toxics Release Inventory. *Environmental Management* 25 (2): 115–127.

Gerber, Brian J. 2002. Administering Environmental Justice: Examining the Impact of Executive Order 12898. *Policy and Management Review* 2 (1): 41–61.

Gianessi, Leonard P., Henry M. Peskin, and Edward Wolff. 1979. The Distributional Effects of Uniform Air Pollution Policy in the United States. *Quarterly Journal of Economics* 91:654–674.

Golden, Marissa Martino. 1998. Interest Groups in the Rule-Making Process: Who Participates? Whose Voices Get Heard? *Journal of Public Administration and Theory* 8 (2): 245–270.

Goldstein, B. D. 1995. The Need to Restore the Public Health Base for Environmental Control. *American Journal of Public Health* 85 (4): 481–483.

Goulder, Lawrence H., and Ian W. H. Parry. 2008. Instrument Choice in Environmental Policy. *Review of Environmental Economics and Policy* 2 (2): 152–174.

Government Accountability Office (GAO). 2005. *Environmental Justice: EPA Should Devote More Attention to Environmental Justice When Developing Clean Air Rules*, GAO-05-289 (July 2005).

Greenstone, Michael. 2002. The Impacts of Environmental Regulations on Industrial Activity: Evidence from the 1970 and 1977 Clean Air Act Amendments and the Census of Manufactures. *Journal of Political Economy* 110:1175–1219.

Greenstone, Michael. 2003. Estimating Regulation-Induced Substitution: The Effect of the Clean Air Act on Water and Ground Pollution. *American Economic Review* 93 (2): 442–448.

Gupta, Shreekant, George Van Houtven, and Maureen L. Cropper. 1996. Paying for Permanence: An Economic Analysis of EPA's Cleanup Decisions at Superfund Sites. *Rand Journal of Economics* 27 (3): 563–582.

Guzy, Gary S. 2000. EPA Statutory and Regulatory Authorities under Which Environmental Justice Issues May Be Addressed in Permitting. Memorandum dated December 1, 2000. http://www.epa.gov/environmentaljustice/resources/policy/ej_permitting_authorities_memo_120100.pdf.

Hamilton, James T. 1993. Politics and Social Costs: Estimating the Impact of Collective Action on Hazardous Waste Facilities. *Rand Journal of Economics* 24 (1): 101–125.

Hird, John A. 1990. Superfund Expenditures and Cleanup Priorities: Distributive Politics or the Public Interest? *Journal of Policy Analysis and Management* 9:455–483.

Kreger, Mary, Katherin Sargent, Abigail Arons, Marion Standish, and Claire D. Brindis. 2011. Creating an Environmental Justice Framework for Policy Change in Childhood Asthma: A Grassroots to Treetops Approach. *American Journal of Public Health* 101 (S1): S208–S216.

Lazarus, Richard J., and Stephanie Tai. 1999. Integrating Environmental Justice into EPA Permitting. *Ecology Law Quarterly* 617:617–678.

Maguire, Kelly, and Glenn Sheriff. 2011. Comparing Distributions of Environmental Outcomes for Regulatory Environmental Justice Analysis. *International Journal of Environmental Research and Public Health* 8:1707–1726.

McCubbins, Matthew D., Roger G. Noll, and Barry R. Weingast. 1989. Structure and Process, Politics and Policy: Administrative Arrangements and the Political Control of Agencies. *Virginia Law Review* 75 (2): 431–482.

Morello-Frosch, Rachel A. 2002. Discrimination and the Political Economy of Environmental Inequality. *Environment and Planning. C, Government & Policy* 20:477–496.

Muller, Nicholas, Daniel Tong, and Robert Mendelsohn. 2009. Regulating NOx and SO2 Emissions in Atlanta. *B.E. Journal of Economic Analysis & Policy* 9 (2). ISSN (Online) 1935–1682, DOI:10.2202/1935-1682.1954, March 2009.

National Environmental Justice Advisory Council (NEJAC). 2002. Integration of Environmental Justice in Federal Programs. http://www.epa.gov/compliance/ej/resources/publications/nejac/integration-ej-federal-programs-030102.pdf.

National Environmental Justice Advisory Council (NEJAC). 2010. National Environmental Justice Advisory Council Meeting. January 27–29, 2010. Meeting transcript from January 28, 2010. http://www.epa.gov/compliance/ej/resources/publications/nejac/nejacmtg/nejac-meeting-trans-012810.pdf.

National Environmental Justice Advisory Council (NEJAC). 2011a. Enhancing Environmental Justice in EPA Permitting Programs. http://www.epa.gov/compliance/ej/resources/publications/nejac/ej-in-permitting-report-2011.pdf.

National Environmental Justice Advisory Council (NEJAC). 2011b. NEJAC Comments to EPA Plan EJ 2014. http://www.epa.gov/compliance/ej/resources/publications/nejac/plan-ej-2014-comments-0511.pdf.

Noonan, Douglas S. 2008. Evidence of Environmental Justice: A Critical Perspective on the Practice of EJ Research and Lessons for Policy Design. *Social Science Quarterly* 89 (5): 1154–1174.

Nweke, Onyemaechi C. 2011. A Framework for Integrating Environmental Justice in Regulatory Analysis. *International Journal of Environmental Research and Public Health* 8: 2366–2385.

Nweke, Onyemaechi C., et al. 2011. Symposium on Integrating the Science of Environmental Justice into Decision-Making at the Environmental Protection Agency: An Overview. *American Journal of Public Health* 101 (S1): S19–S26.

O'Neil, Sandra George. 2007. Superfund: Evaluating the Impact of Executive Order 12898. *Environmental Health Perspectives* 115:1087–1093.

Parry, Ian W. H. 2004. Are Emissions Permits Regressive? *Journal of Environmental Economics and Management* 47:364–387.

Ringquist, Evan J. 2005. Assessing Evidence of Environmental Inequities: A Meta-Analysis. *Journal of Policy Analysis and Management* 24 (2): 223–247.

Ringquist, Evan J. 2011. Trading Equity for Efficiency in Environmental Protection? Environmental Justice Effects from the SO2 Allowance Trading Program. *Social Science Quarterly* 92 (2): 297–323.

Rubin, Scott J., and Robert Raucher. 2010. Applying Environmental Justice to Drinking Water Regulation. White paper for the National Rural Water Association. http://www.nrwa.org/benefits/whitepapers/TOC.htm.

Schmalensee, Richard, and Robert N. Stavins. 2013. The SO2 Allowance Trading System: The Ironic History of a Grand Policy Experiment. *Journal of Economic Perspectives* 27 (1): 103–122.

Shapiro, Marc D. 2005. Equity and Information: Information Regulation, Environmental Justice, and Risks from Toxic Chemicals. *Journal of Policy Analysis and Management* 24 (2): 373–398.

Shimshack, Jay P., and Michael B. Ward. 2010. Mercury advisories and household health trade-offs. *Journal of Health Economics* 29:674–685.

Sigman, Hilary. 2001. The Pace of Progress at Superfund Sites: Policy Goals and Interest Group Influence. *Journal of Law & Economics* 44:315–344.

Turaga, Rama Mohana R., Douglas S. Noonan, and Ann Bostrom. 2011. Hot Spots Regulation and Environmental Justice. *Ecological Economics* 70 (7): 1395–1405.

5

Evaluating Environmental Justice: Analytic Lessons from the Academic Literature and in Practice

Ronald J. Shadbegian and Ann Wolverton[1,2]

Economists have long concerned themselves with the efficiency and distributional implications of environmental policy. Studies of efficiency quantify the net change in welfare (total benefits minus total costs), expressed in dollar terms. In contrast, distributional analysis concentrates on understanding how the potential costs and benefits of cleaning up the environment vary based on income (e.g., Freeman 1969; Freeman 1972; Asch and Seneca 1978), though a few early studies also considered race (e.g., Berry 1977). While a study of environmental justice also examines the way in which the outcomes of environmental policy are distributed across various populations, it differs from traditional distributional analysis (or at least an economist's view of it) in a number of ways. First, as envisioned in Executive Order 12898 (EO 12898), environmental justice focuses on particular population groups—specifically, minority, low-income, and indigenous populations—while a distributional analysis discusses how changes in welfare vary by income or education but rarely contemplates how effects vary by race or ethnicity. Second, per the executive order, an environmental justice analysis is concerned with "disproportionately high and adverse human health or environmental effects," but it does not usually discuss monetized values of costs and benefits.

Recently, executive branch agencies have begun to more regularly consider environmental justice in their analyses for rulemakings, with the Environmental Protection Agency (EPA) being at the forefront of this effort. In 2010, Administrator Lisa Jackson listed environmental justice as one of the EPA's top priorities, and the analysis of the potential environmental justice effects of rulemakings substantially increased under her leadership. Specifically, there has been a "dramatic increase in the number of environmental justice analyses of EPA rules between 2010 and 2012" from less than two per year on average between 1995 and 2009 to an average of more than twenty analyses of environmental justice concerns conducted annually between 2010 and 2012 (EPA 2013a). With this increased attention on evaluating potential environmental justice concerns associated with rulemaking comes the necessity of grappling with a number of analytic issues raised in the academic literature.

In this chapter, we highlight several of these analytic issues, how they have been treated in the academic literature, and what approaches have been taken to address them in practice. The first section summarizes the history of environmental justice analysis for rulemakings at the EPA. The next section discusses five major analytic issues that are relevant when evaluating the effects of a regulation on minority or low-income populations and how they have been approached in the academic literature compared to a sample of recent environmental justice analyses at the EPA. Finally, the last section highlights several areas for future research related to analytic issues that are recognized in the literature and by the EPA, but are often quite challenging to address in practice.

History of Analysis of Environmental Justice at the EPA

At least partially motivated by the "stylized fact" derived from many academic studies of a correlation between emissions or environmental risk and low-income or minority households (Banzhaf 2011), President Bill Clinton issued EO 12898 on February 11, 1994. The EPA created an Office of Environmental Justice even earlier, in 1992, to ensure "the fair treatment and meaningful involvement" of individuals in the development, implementation, and enforcement of its regulations and policies. The EPA also includes the goal of ensuring "the same degree of protection from environmental and health hazards" in its definition of environmental justice (www.epa.gov/environmentaljustice). Banzhaf (2011) notes that analysis of a regulation's distributional implications becomes even more important in the context of this expanded definition.

Analysis of a regulation's impacts is a key input into the EPA's decision-making process. Several options for regulating a pollutant are iteratively discussed and analyzed within the agency, including the costs, benefits, and other impacts of each option when feasible. Based on this information, a proposed rulemaking is published in the Federal Register and accompanying analyses are available in the docket for public comment. Information collected through the public comment process is then considered and potentially integrated into the rulemaking—both in terms of how the pollutant is regulated and the data and methods used to analyze its impacts—and published as part of the final rulemaking package. The Office of Management and Budget's Circular A-4 provides general guidance on the conduct of regulatory analysis, with an emphasis on estimation of costs and benefits, but it also recommends a separate description of distributional effects. While Circular A-4 does not specifically refer to environmental justice, it notes that important distributional effects of regulatory options should be described "quantitatively to the extent possible, including the magnitude, likelihood, and severity of impacts on particular groups" (OMB 2003).

EPA interim guidance on environmental justice in the regulatory process (EPA 2010) also directs analysts to evaluate whether environmental justice is likely to be an issue for the regulation in question. For instance, an analytic blueprint is developed early in the regulatory process. In it, analysts are expected to "articulate potential environmental justice concerns and how [they] will explore and approach them in developing the action," including the identification of "available EJ assessment tools, as well as related needs for data collection, expertise, and resources" and "potential analytical issues that will need to be raised to management or otherwise addressed" (EPA 2010). Any analysis of environmental justice issues is then integrated into discussions with decision makers prior to the proposal and finalization stages of the rulemaking (e.g., during "options selection" with senior management).

While the EPA has worked to integrate environmental justice into its deliberative processes, it has been criticized for not incorporating environmental justice more fully into economic analyses of rulemakings (GAO 2005; Vajjhala et al. 2008; Banzhaf 2011). For instance, the GAO (2005) reviews three air rules—selected because they mentioned potential environmental justice concerns when proposed—and finds that "after reviewing [public] comments [on the issue], the EPA did not change its final economic reviews to discuss [the] rules' potential environmental justice impacts." Using content analysis, Vajjhala et al. (2008) find that the EPA regulatory impact analyses (RIAs) analyzed include some environmental justice–related terms and that their usage has increased over time. However, acknowledgment of possible population-level disparities in impacts is limited. Banzhaf (2011) states that the EPA tends to issue "perfunctory, pro forma assertions that it is not creating or exacerbating an environmental injustice" in its RIAs without providing evidence that this is the case.

At the time of the GAO report, analysts had available to them general guidance from two EPA publications on how to evaluate environmental justice issues for rulemakings: the *Guidelines for Preparing Economic Analyses* (EPA 2000), and the *Toolkit for Assessing Potential Allegations of Environmental Injustice* (EPA 2004). We refer to these documents in this chapter as the *Economic Guidelines* and *EJ Toolkit*. The *Economic Guidelines* briefly outlined a broad framework for conducting distributional analysis for economically significant rulemakings,[3] which included identifying potential areas in need of study, defining key distributional variables and relevant equity dimensions, and then measuring the equity consequences of the rule. The discussion did not include detailed information on how to actually evaluate potential environmental justice concerns—for instance, it did not discuss or recommend particular analytic tools or methodological challenges. Furthermore, the *Economic Guidelines* addressed a range of distributional issues and did not solely focus on environmental justice issues. The *EJ Toolkit* (EPA 2004) was designed to investigate "allegations of environmental injustice." While an investigation could encompass a national rulemaking, the *EJ Toolkit* was mainly geared toward evaluating allegations at a particular site or location (also see Banzhaf 2011). It recommends a qualitative screening-level assessment based on readily available information to determine whether further analysis is warranted and, if a potential concern is identified, progressing to a more refined quantitative analysis.

More recently, the EPA has been developing detailed technical guidance to aid analysts in the evaluation of environmental justice issues for rulemakings.[4] The guidance aims to increase consistency across analyses while maintaining flexibility in approach to allow for "data limitations, time and resource constraints, and analytic challenges" (EPA 2013b). It includes questions for analysts to ask when assessing environmental justice issues in the baseline (i.e., what the world would look like in the absence of regulation) and for each policy option under consideration, as well as detailed discussions of methods, data, and other tools available to potentially answer them. Recommended "best practices" when conducting an environmental justice analysis include carefully selecting the comparison group against which environmental or health impacts on the basis of race or income are measured, characterizing the distribution of risks, exposures, or outcomes across individuals, life stages, gender, or other relevant categories within each population group when feasible, and conducting sensitivity analysis for key assumptions or parameters that may affect findings (among others). The draft technical guidance also offers advice on how to integrate environmental justice into the planning and scoping stages of a human health risk assessment and alerts the analyst to potential measurement and methodological considerations. The EPA also recently released a new chapter of its *Economic Guidelines* that discusses data and methodological choices when evaluating environmental justice concerns for economically significant rules (EPA 2014a).

Lessons from the Academic Literature

Early environmental justice literature was a direct response to media and activist attention on the siting decisions of landfills and hazardous waste sites. While these papers used small data sets and relatively simple empirical techniques (e.g., pairwise correlations that do not control for confounding factors), they are notable because they point out the potential for a positive correlation between the incidence of pollution and poor or minority populations (e.g., Bullard 1983, GAO 1983, United Church of Christ 1987). When accounting for other factors, more recent studies find mixed evidence for the role that race and income play.[5] For instance, Kriesel et al. (1996) find that poverty and race are positively related to emissions, but when education and property values are added to the regression, race is no longer significant. Gray and Shadbegian (2004) find that pulp and paper plants emit more pollution in poor areas, but less in minority areas. Examining plant siting decisions in Texas, Wolverton (2009) finds that input-related cost factors are more important than race and ethnicity, while poverty discourages siting. Morello-Frosch, Pastor, and Sadd (2001) find that race is positively correlated with lifetime cancer risk from cumulative exposure to outdoor hazardous air pollutants, but income is negatively correlated with this risk. Using atmospheric dispersion modeling, Ash and Fetter (2004) find that minority and low-income households are exposed to greater concentrations of air pollution, while Depro and Timmins (2012) find that minorities face "very different (and disadvantageous) tradeoffs" between housing services and ozone concentration in the San Francisco area compared to whites.[6,7]

As the academic environmental justice literature has grown in breadth and sophistication, several analytic issues have been raised that are relevant when predicting ex-ante impacts of regulation by population group. For example, the review by Mohai and Bryant (1992) indicates that results may be sensitive to geographic scope, variation in neighborhood definition (spatial scale), empirical technique, control variables, and type of facility examined. However, academic studies differ from analyses to inform rulemaking in a number of key ways. First, academic studies of environmental justice often focus on a particular city or state and are ex-post assessments.[8] Environmental justice analyses to inform federal regulatory decisions are typically conducted at the national level and are ex-ante assessments of incremental effects.[9] Another key difference is that academic studies of environmental justice rarely concern themselves with what effects are directly attributable to a specific policy or action, instead focusing on the general relationship between location, emissions, or risk and vulnerable communities (see also the discussion in chapter 4). However, policy attribution is of paramount importance when examining whether a particular regulation will mitigate or exacerbate any existing environmental inequities.

In this discussion, we examine several analytic issues that are regularly discussed in the literature and that are of particular importance to EPA analysts performing environmental justice analyses. More specifically, we cover the geographic scope and scale of the analysis, the identification of potentially affected populations, the selection of a comparison group, how to spatially identify effects on population groups, and how exposure or risk is measured in an analysis. For each issue, we examine how it is treated in practice by referring to five proposed or final EPA rulemakings across a variety of media (i.e., air, water, and solid waste), published between 2008 and 2012 (see table 5.1). It is important to keep in mind a major caveat to this approach: these examples may or may not be representative of the larger set of environmental justice analyses conducted by the EPA or federal agencies more generally.

Table 5.1

Sample of Recent EPA Rulemakings that Evaluate Potential Environmental Justice (EJ) Concerns

Rulemaking/ Statute	Status/ Publication Year	Summary and Potential EJ Considerations
MATS, Clean Air Act (CAA)	Final; 2011	• Limits emissions of HAPs from electric utilities. • Because of possible shifts in electricity generation from utilities with lower emission rates to those with higher emission rates, emissions could increase in some communities.
Revised DSW Rule, RCRA	Proposed; 2011	• Allows the recycling of hazardous secondary materials. • The EPA originally concluded that the rule would preclude any increase in risk to communities near these facilities. • After rule publication, the EPA received comments that some communities may be negatively affected when facilities process more hazardous material than before. The EPA performed an expanded EJ analysis.
National Pollutant Discharge Elimination System (NPDES) CAFO Reporting Rule, CWA	Proposed, then withdrawn; 2011	• Intended to improve the general public's access to information on large CAFOs. • While the proposed regulation would allow all individuals easier access to emissions information for CAFOs, the EPA used its EJ analysis to identify communities in need of additional outreach.
Lead RRP Rule, Toxic Substances Control Act (TSCA)	Final; 2008	• Requires the renovators of apartments, pre-1978 houses, day-care centers, and kindergartens to follow work practices to reduce lead dust. • Since low-income households are more likely to reside in older or rental housing, they are more likely to face new requirements. • If households delay or avoid renovation due to additional costs, they also do not reap the benefits of reducing lead exposure.
NESHAP: Hard chromium electroplating operations, CAA	Final; 2012	• Sets maximally achievable control technology air emission standards for electroplaters. • Characterizes how potential cancer risks vary with demographic composition as part of a residual risk and technology review.

Geographic Scope

One reason why studies offer such mixed evidence that environmental impacts vary by race and income may be differences in the scope of the analysis (Bryant and Mohai 1992). Some studies are national in scope, while others focus on a particular urban area or region. Results from state- or city-specific analyses cannot be easily generalized to other geographic contexts, while national studies could potentially disguise the importance of some socioeconomic factors.

Empirical evidence indicates that results are sensitive to geographic scope in some cases but not others. For instance, Zimmerman (1993) finds that, at both the national and regional levels, a greater percentage of minorities live near inactive hazardous waste sites on the National Priorities List (NPL), but the percentage who live in poverty is insignificant. Likewise, Anderton et al. (1994) find that minority or poor neighborhoods do not host a statistically different number of hazardous waste facilities than other neighborhoods at the national or regional level, though Davidson and Anderton (2000) find some notable differences in the location of waste facilities by race and income across urban and rural areas. Bowen et al. (1995) find that emission releases and minority populations are highly correlated at the state level, but not at the metropolitan level. The authors note that a state-level analysis is likely less appropriate in this case because both the industry studied and the minority populations in the state are concentrated in the same metropolitan area. To make sense of potentially contradictory results across studies, Ringquist (2005) uses a meta-analysis to compare 49 environmental justice studies. He finds that studies that are national in scope tend to find smaller differences in impacts by race than studies limited to a particular region, state, or municipality, but that the scope of the analysis does not seem to explain differences in impacts for low-income households across studies.

Baden et al. (2007) review over 100 environmental justice studies and also find that "significant findings are not independent of scale." However, they note that there has been little systematic exploration of the influence of scope on the results *within* a study. Often, authors vary several factors together, making it difficult to parse their separate effects on the results. For this reason, in a study of Superfund sites, Baden et al. (2007) vary the scope while holding other aspects constant. They find that race and ethnicity are positively correlated with the presence of a Superfund site at the national and state levels for many different spatial scales (and actually grow in magnitude and significance at the state level), but these relationships are no longer significant for race at the county level. Income is significant and negatively correlated with a Superfund site at the national and county levels, but not at the state level. Wolverton (2012) examines whether any differences by race or income in polluting firms' location decisions are robust to scope, all else being constant. While the results for race and income are consistent across all specifications, poverty is somewhat sensitive to geographic scope (i.e., city versus state).

In Practice While the academic literature varies widely in its choice of scope, most EPA rulemakings apply nationally. For this reason, it follows that the environmental justice analyses conducted by the agency are often national in scope. For instance, the analyses of environmental justice issues conducted for the Mercury and Air Toxics Standards (MATS), Concentrated Animal Feeding Operations (CAFOs) Reporting, Hard Chromium Electroplating National Emissions Standards for Hazardous Air Pollutants (NESHAP), and Lead Renovation, Repair, and Painting (RRP) rules all compare changes in risk, exposure, or emissions to the national average for other communities. The choice of national scope makes sense in these instances since what is being analyzed are the effects of a regulation that will apply to all affected facilities in the United States. However, in light of the varying results in the academic literature with regard to geographic scope, there may be cases where it makes sense to also provide more disaggregated comparisons for a national analysis. If a regulation applies to facilities that are concentrated geographically, then a study that is national in scope may not be necessary to characterize its main impacts. Likewise, differences in the way federal programs are implemented or enforced by the regions or states may justify additional analysis at a subnational level. For instance, in addition to analysis that is national in scope, the expanded analysis of the 2008 Definition of Solid Waste (DSW) final rule included state-level comparisons to account for differences in the way Resource Conservation and Recovery Act (RCRA) programs are implemented at the state level.

Identifying Potentially Affected Populations

The academic environmental justice literature has traditionally focused on low-income and minority communities (effects on indigenous communities are rarely evaluated), but researchers have used a variety of ways to define these population groups.[10] In some cases, minority is defined as nonwhite (e.g., Hamilton 1995; Gray and Shadbegian 2004, 2007; Wolverton 2009), while in other cases, it is defined more narrowly to focus on a particular race or ethnicity—for instance, African-American, black, or Hispanic (e.g., Been and Gupta 1997; Sadd et al. 1999; Morello-Frosh et al. 2001; Baden et al. 2007). To examine impacts on low-income communities, studies have used the poverty threshold or some multiple of it to differentiate this population group from others (e.g., Hamilton 1995; Arora and Cason 1998; Gray and Shadbegian 2004, 2007; Wolverton 2009). In some cases, level of education has been used as a proxy for low income (e.g., Hamilton 1995, Arora and Cason 1998). While this variation makes it somewhat challenging to compare results across studies, the way in which the terms *minority* and *low income* are defined and measured seems contingent on data availability and an assessment of the most relevant way to capture the particular subset of individuals likely to be most affected.

In Practice Similar to the academic literature, in the rulemakings that we examine, the EPA defines potentially affected population groups in various ways to evaluate environmental justice concerns. For instance, to assess the impact of changes in mercury exposure associated with MATS, the EPA focuses on population groups with an especially high potential risk of exposure due to high rates of fish consumption: Laotian subsistence fishers, low-income African-American and white recreational and subsistence fishers in the Southeast, Chippewa/Ojibwe members in the Great Lakes area, low-income female recreational/subsistence fishers, and Hispanic subsistence fishers.

For the hard chromium electroplating NESHAP, minority and low-income populations are defined in several ways: minority as well as African-American, Native-American, and Hispanic; households living below the poverty line, and those earning less than median U.S. household income. In addition, the EPA examines the distribution of potential cancer risk associated with the hard chromium electroplating NESHAP by age and education. For the expanded-DSW analysis, the EPA focuses on the populations specifically mentioned in EO 12898—minority, Native American and Alaska Native, and people living in poverty—as well as children under the age of five. These demographic groups were chosen based on a petition from the Sierra Club, comments received in response to the petition and public meetings to discuss the 2008 DSW final rule and environmental justice analysis, as well as a review of previous environmental justice analyses. For the CAFOs Reporting and Lead RRP rules, the EPA examines where minority and low-income populations are located. No information is given in the CAFOs Reporting rule for how these groups are defined. The Lead RRP rule defines *low income* as households living below the poverty threshold, while *minority* indicates individuals who are black/African-American alone or Asian alone. Like the expanded-DSW analysis, the Lead RRP rule also includes an analysis for children—which Lead RRP defined as those under the age of 6—by "non-parental arrangement," poverty status, and race.[11]

Selecting a Comparison Group

To determine whether impacts differ by race or income, one needs to compare incremental impacts experienced by one group to those of another. However, there has been little explicit study or discussion of how the comparison group may affect the results of an environmental justice study.

Bowen (2001) notes that limiting the comparison group to the same metropolitan area as those affected by a polluting site may be more defensible than a national comparison group because of regional variation in level of economic growth, industry concentration, and socioeconomic characteristics. Davidson and Anderton (2000) compare neighborhoods with a RCRA waste facility to those without such a facility. However, because they are concerned about the possibility of "spurious effects of urban-rural residential differences," they also use neighborhoods without a waste facility but within the same MSA or nonmetropolitan area (in the case of rural areas) as a comparison group. Ash and Fetter (2004) use urban-area fixed effects to sweep out such differences so that they can make a national comparison of exposure across population groups.

Using meta-analytic techniques, Ringquist (2005) observes that studies that limit the comparison group before conducting the analysis—as opposed to controlling for factors that affect environmental exposure or risk directly in a regression—find fewer differences in impacts by race and income. Ringquist argues that restricting the comparison group before conducting analysis based on the presence or absence of emissions—which are also correlated with risk and exposure among poor or low-income groups—may bias the results toward zero or "reduce the power of statistical tests by reducing sample sizes."

In Practice For most of the EPA rulemakings that we examine, the comparison group has been defined as the national average. In a proximity-based analysis, this is often the average for individuals living far enough away from the facilities in question to not be affected by changes in emissions associated with the rulemaking. For instance, the expanded-DSW analysis compares the demographic characteristics of population groups within a particular distance of each facility to average demographic characteristics nationally. In the Lead RRP rule, the EPA compares minority groups by renter or owner status to "white alone," and households below the U.S. poverty threshold to those above the U.S. poverty threshold. For the MATS rule, the EPA compared the change in PM2.5 mortality risk[12] for the population living in "high-risk" counties (i.e., the top fifth percentile of the distribution) with the risk of those living in all other counties.

There are cases where—even though a national comparison is presented—the EPA also provides more disaggregated comparisons. For instance, because CAFOs are located in rural areas, the EPA excludes urban census tracts from the comparison group for the CAFOs Reporting rule analysis: the socioeconomic characteristics of each census tract with a CAFO are compared to those of the average rural census tract. Likewise, the expanded-DSW analysis examines urban and rural populations separately to test whether the tendency for hazardous material recycling facilities to locate in highly populated urban areas affects the environmental justice analysis. As previously mentioned, it also includes state-level comparisons due to differences in RCRA implementation.

While the MATS, hard chromium electroplating NESHAP, CAFOs Reporting, Lead RRP, and DSW rules all compare the demographic characteristics within the affected area to the average characteristics of the comparison group, the expanded-DSW analysis makes two different types of comparisons. The analysis examines whether (1) minority and low-income populations face a substantially higher probability of being affected by a DSW facility compared to nonminority or non-low-income populations, and (2) whether the populations affected by DSW facilities are disproportionately minority, low-income, or both.

How to Spatially Identify Effects on Population Groups

When human health outcomes are spatially distributed, it is necessary to spatially characterize the affected and unaffected areas and then aggregate effects by population group to make comparisons. However, selecting an appropriate geographic unit of analysis is not necessarily a straightforward exercise, as emissions are unlikely to be neatly contained within specified boundaries unless they are defined at a highly aggregate level. In addition, selecting the appropriate level of aggregation raises the "modifiable areal unit problem." An analyst typically has multiple choices of areal units for analyzing aggregated data, and the choice of unit can affect measurement of distributional effects (e.g., Mohai and Bryant 1992; Baden et al. 2007).

The nature of the pollutant is one input into the choice of geographic unit of analysis. For instance, does the pollutant in question travel long distances, affecting individuals many kilometers away, or does it have more localized effects? A second input into the decision of geographic scale is data availability. How spatially resolved are the available data on environmental quality, socioeconomic characteristics, and indicators of exposure or risk? A third input into choice of geographic scale is whether the analyst should use predefined geographic aggregations or build them. Analysts who rely on predefined spatial aggregations often turn to U.S. Census data. While convenient, census-based definitions often use topographical features such as rivers, highways, and railroads as boundaries, which do not necessarily correspond to the affected area. In most instances, if an analyst has access to a fate and transport model or to geographic information system (GIS)–based software, then it may be possible to specify a uniform spatial buffer around an emissions source that more accurately reflects its characteristics.[13]

Whether using a census-based definition or a GIS-based buffer, environmental justice analyses are all subject to the modifiable areal unit problem. This is because the level of aggregation can affect the mean and variance of socioeconomic characteristics that are not uniformly distributed on a spatial scale (Konisky and Schario 2010). For example, if households spatially cluster based on race or income, then a less aggregated spatial scale will likely have less variation in these characteristics than would be true for a larger geographic area (Baden et al. 2007). A finer spatial scale also increases sample size, which can reduce variance. The method of drawing a buffer around an identifiable pollution source is less appropriate for disperse pollutants (Banzhaf 2011). Instead, an analyst may want to first identify the community and then measure its exposure.

Several papers have shown that the way in which communities around a polluting source are characterized may affect the results, though evidence is mixed. For instance in his meta-analysis of 49 environmental justice analyses, Ringquist (2005) finds that studies using smaller geographic units of analysis find larger differences in impacts on the poor than those conducted at a more aggregate level. In contrast, he finds that the geographic unit of analysis does not seem to explain differences across studies in impacts by race. Baden, Noonan, and Turaga (2007) also note a wide variation in the geographic unit of analysis across the more than 100 environmental justice studies they reviewed. They find that "studies at smaller scales (e.g., tract, block group) appear to exhibit more statistically insignificant findings than at larger scales." As previously noted, however, many of the studies included in these reviews simultaneously explore the sensitivity of results to changes in geographic scale and scope of the study. For example, Shadbegian and Gray (2012), who define the potentially affected area as a 1-, 5-, and 10-mile buffer surrounding the site, find the distribution of enforcement activity is not sensitive to geographic scale. Similarly, Konisky and Schario (2010), who rely on census-based definitions (tracts and block groups), also find that their results with regard to enforcement are largely robust to geographic scale.

Chakraborty and Maantay (2011) note that some regression techniques are predicated on assumptions that are violated by spatial data used in many environmental justice studies. For instance, spatial clustering violates the assumption that error terms are independently and identically distributed. In these instances, spatial econometric techniques are required. However, they are rarely used in the environmental justice literature. Two exceptions are Gray and Shadbegian (2007) and Chakraborty (2011), who employ spatial econometric techniques to examine environmental justice. Gray and Shadbegian (2007) find, whether they use spatial econometric techniques or not, that polluting plants in poor and nonminority neighborhoods do not have significantly worse environmental performance. On the other hand, Chakraborty (2011) finds that ignoring spatial autocorrelation leads to potentially invalid conclusions regarding the association of race with increased lifetime cancer risk.

In Practice The five environmental justice analyses that we examine are for regulations where emissions and their effects are transmitted spatially via air or water. As with the academic literature, the EPA has used a variety of methods for spatially identifying the effects of changes in exposure or risk on population groups. This variation reflects data constraints, but also an evaluation of the geographic scale that comes closest to mimicking the emission patterns associated with a particular type of pollutant.

In the MATS, DSW, and hard chromium electroplating NESHAP rules, the EPA defines an affected area based on GIS-defined buffers. The MATS analysis uses a 5-kilometer radius because air quality modeling indicated that an individual polluting source generally has the highest ambient air levels for the primary hazardous air pollutants (HAPs) emitted by electric utilities within this distance. The expanded-DSW analysis uses a 3-kilometer radius around each DSW facility to define the affected community because it approximates the likely area that could be affected by a dangerous event such as a fire or explosion. The hard chromium electroplating NESHAP examines the cancer risks within a 50-kilometer radius (built up from census block data) based on air dispersion modeling.

In other cases, the EPA has relied on census-based definitions. For instance, for MATS, the EPA relies on the county designation as the geographic scale since this is the smallest scale at which baseline mortality rates are available. The CAFOs Reporting rule examines the density of CAFOs by county and their proximity to minority or low-income populations at the census tract level. For the Lead RRP rule, data from the 2000 Census yield some information on socioeconomic and housing characteristics. However, they are presented at only a national scale in broad summary statistics. No further quantitative analysis was deemed feasible, which made the decision of the most appropriate spatial unit of aggregation moot.

How Exposure or Risk Is Measured

If data availability were not an issue, the best way to capture differences in health effects would be to use information on variation in exposure and risk by race and income to characterize differences across population groups. However, these types of data are often quite limited. A common surrogate for direct measures of risk or exposure used in the literature is proximity, or the distance an individual or group is from a pollution source (e.g., UCC 1987; Been 1994; Baden and Coursey 2002; Wolverton 2009; Cameron, Crawford, and McConnaha 2012). Of the more than 100 studies that Baden et al. (2007) review, about 61 percent are proximity-based analyses. Studies that rely on emissions or risk data often focus on air pollution (e.g., Kriesel et al. 1996; Arora and Cason 1998; Gray and Shadbegian 2004; Morello-Frosh, Pastor, and Sadd 2001).

Many studies in the literature examine pollutant sources particularly amenable to proximity-based approaches, such as hazardous waste sites, Superfund sites, or landfills. That said, a proximity-based approach assumes that the effects of the pollutant occur only within a designated boundary, that all individuals residing within the boundary are equally exposed, and that the number of sources does not matter (e.g., Mohai and Saha 2006; Chakraborty, Maantay, and Brender et al. 2011; Banzhaf 2011). A count regression technique, such as a Poisson regression or a negative binomial distribution, can improve on a logistic regression approach by accounting for the number of sources of pollution within a neighborhood (e.g., Wolverton 2009, 2012), although it still cannot control for differential effects in exposure. A regression that weights the effect on a neighborhood by distance may be able to partially account for differences in exposure in some cases (e.g., Mohai and Saha 2006). Nevertheless, this approach still cannot account for any differences in susceptibility across population groups due to genetic differences, health and nutritional deficits, or varying economic conditions, which would require population group specific concentration-response functions. However, Banzhaf (2011) notes that there is rarely enough information available in the literature to specify concentration-response functions for particular population groups.

In Practice In the environmental justice analyses that we examine, the EPA identifies potential effects of a regulation by race and income through a qualitative assessment, the use of maps, proximity-based analysis, and a combination of air quality modeling and health impact analysis when proximity-based analysis would serve as a poor proxy for exposure and risk. The EPA characterizes preexisting (or baseline) differences in risk or exposure in the five analyses we examine. Given the tools and data available (for instance, proximity-based analysis), the ability to differentiate effects by regulatory option is often not feasible.

The expanded-DSW environmental justice analysis uses a proximity-based approach to evaluate the potential for environmental justice concerns. Since analysts do not have information on the extent to which the regulation will change hazardous waste recycling at each facility and the risks that this may impose on nearby households, distance from a facility is used as a proxy for the risk of living near a facility. The analysis also qualitatively discusses factors that may affect the susceptibility of affected populations to pollutants associated with nearby DSW facilities. MATS also utilizes a proximity-based approach as a proxy for exposure to HAP emissions from electric utilities because the health effects, which include elevated lifetime cancer risks in nearby communities, are expected to be localized.

In addition to the proximity-based approach for MATS, the EPA performs an analysis to characterize the change in the distribution of mortality risk associated with PM2.5 after implementation of the MATS final rule. Because criteria air pollutants can travel long distances from their original sources and the formation of PM2.5 is governed by complex nonlinear atmospheric chemistry, a proximity-based analysis is deemed inappropriate. Instead, the EPA stratifies baseline mortality rates by race and uses mortality risk coefficients that vary by education level to analyze the change in mortality risk associated with PM2.5 across population groups.[14] The EPA also includes a qualitative assessment of the effects of the rule on population groups with an especially high potential risk of mercury exposure. In this case, due to data constraints, the analysis reviews the literature to identify the population groups likely to be at high risk for mercury exposure based on consuming more fish than the daily average in the general population. The analysis then uses projections of county-level growth to estimate the number of people in each population group at risk of mercury exposure in 2016 without MATS (i.e., in the baseline) and presents this information in a series of maps.

For the hard chromium electroplating NESHAP, there is enough information available to estimate risk. The EPA uses air dispersion and human exposure modeling to estimate baseline cancer risks due to the inhalation of HAPs from nearby hard chromium electroplating facilities. Inhalation risk is estimated near the centroid of each census block and then applied equally to each individual living in the block by population group. The EPA then uses this information to calculate the average risk and the distribution of risk within a population group.

For the other rules we discuss, far less information is available to evaluate the effects of the rule by population group. For the CAFO Reporting and Lead RRP rules, information on baseline characteristics is summarized in tables or maps. Data constraints prevent the EPA from providing a quantitative estimate of how the proposed requirements of the CAFOs Reporting Rule could affect emissions from CAFOs operations located near minority or low-income populations. Instead, the EPA uses maps to offer insights into the preexisting distribution of CAFOs operations. The location of farms with a large number of animals (excluding beef cattle) is used as a proxy for the location of large CAFOs. The EPA calculates and maps the density of CAFOs by county. Census-tract-level data on the location of communities with a high density of minority or low-income populations are overlaid with these data to produce a map showing areas with both a high density of CAFOs and high concentrations of low-income or minority populations in rural areas in the United States.

The Lead RRP rule presents some basic quantitative information but is largely qualitative. For instance, it presents statistics categorizing the count and percentage of households living below the poverty threshold by the year the house of residence was built and whether it is rented or owned. Likewise, it presents the count and percent by renter or owner, though not by housing vintage, which could be important since homes built prior to 1978 likely contain lead paint. In addition, the EPA discusses several reasons why low-income or minority individuals may be more susceptible to lead exposure. For instance, lack of access to a healthy diet or to enough calories can heighten one's vulnerability to the negative health effects of lead dust.

Areas in Need of Further Research and Emerging Analytic Issues

The EPA has increasingly integrated environmental justice concerns into its assessments of the effects of regulations in recent years. However, analysis of environmental justice in the context of rulemaking raises a number of analytic issues, which we are discussing in this chapter: geographic scope and scale of the analysis, defining potentially affected populations, selecting a comparison group, spatially identifying effects on population groups, and how exposure or risk is proxied. For each issue, we have described its treatment in the academic literature and in five recent proposed or final EPA rulemakings. In general, we find that the EPA has used approaches to address each analytic issue that are widely utilized in the academic literature. However, a key challenge for the EPA is data constraints for evaluating the effects of national level policy, which has meant that the agency has relied on qualitative assessment in several cases and has restricted itself to an evaluation of baseline conditions for all five of the rulemakings that we discuss in this chapter.

In addition to what we discuss here, the literature has raised several issues that are quite challenging to address in practice. For example, are there evaluation methods that can incorporate both efficiency and equity considerations? How should costs be considered when evaluating environmental justice concerns? And, how can we improve the way we measure risks associated with environmental exposure? These areas are ripe for future research. Given the EPA's increased emphasis on environmental justice issues in rulemaking, these issues also warrant careful thought regarding when and how they may be applied in a national regulatory setting.

Efficiency and equity considerations may not always line up well directionally for a given policy. For instance, a cost-benefit analysis could indicate that a policy is welfare-improving, meaning that total quantified benefits exceed costs, but this is unlikely to hold true for every individual or firm affected by the policy. A policy may raise environmental justice concerns when this does not hold true for particular population groups. Efficiency and equity are both important considerations for shaping environmental policy, but currently there is not a clear way to combine them into an analysis that would explicitly account for and weigh them. The welfare theory of economics discusses the notion of specifying an aggregate welfare function for society, but how individuals are weighted within it is ultimately a societal decision, not one that economics alone can answer. Exploratory work by Cropper (2010) indicates that individuals have a preference for a more equitable distribution than one that is less equal. This work points to the possibility of incorporating individual preferences for equity into a willingness-to-pay measure, setting aside whether there are also societal preferences. Current EPA guidance (EPA 2013b) directs analysts to provide information to decision makers on notable differences in impacts across population groups by race or income but leaves it to the decision makers to determine, on a case-by-case basis, how to weigh environmental justice concerns against a myriad of other factors. Several recent studies also examine the potential merits and challenges of using inequality indices in such an exercise (Maguire and Sheriff 2011, Harper et al. 2013).

A distributional analysis evaluating the net effects of a policy on welfare would also account for the incidence of costs. However, as already noted, this has not traditionally been a consideration in environmental justice analyses, either in academia or in practice. Banzhaf (2011) points out that a narrowly defined assessment of "equity could result in counterintuitive and unintended, even perverse, decision rules for policy." Future research could investigate when the distribution of costs may be relevant, and in particular how an assessment of environmental justice might change once cost considerations are incorporated. As discussed in this chapter, the EPA qualitatively considered the cost implications on the basis of race and income for only one regulation in our sample.

It is also important to continue to make progress in the measurement of risk. In many cases, researchers and practitioners have relied on rough approximations of risk due to lack of direct measures. For instance, several of the regulations discussed in this chapter rely on proximity-based analysis or a qualitative discussion to assess differences in risk. In addition, the ability to differentiate risk on the basis of race, ethnicity, or income is often limited by a lack of underlying epidemiological data. Finally, while a rich methodological and policy literature exists, our ability to assess cumulative risk is often limited due to data, methodology, or other constraints. While California recently began to develop a "screening methodology for evaluating the cumulative impacts of multiple sources of pollution in specific communities or geographic areas" (California EPA 2010), this is described as only a first step to generate scientific and technical discussion and should not be used for regulatory purposes. Additional research into these areas could provide analysts with better tools to investigate the environmental justice implications of policy going forward.

Notes

1. Both authors are employed by the U.S. Environmental Protection Agency (EPA) in the National Center for Environmental Economics (NCEE). Questions or comments on this chapter can be directed to Shadbegian.ron@epa.gov or Wolverton. ann@epa.gov.

2. Any opinions and conclusions expressed herein are those of the authors and do not necessarily represent the views of the EPA.

3. The *Economic Guidelines* apply to economically significant rules, those with benefits or costs above $100 million in any one year.

4. A draft was released for public comment in 2013 (http://www.epa.gov/environmentaljustice/plan-ej/rulemaking.html) and is under review by the EPA's Science Advisory Board. It discusses many of the analytic issues highlighted in section 3 of this chapter, "Lessons from the Academic Literature." The EPA expects it to be finalized by the end of 2014 (EPA 2014b).

5. Also, see Zimmerman (1993), Been (1994), Hamilton (1995), Been and Gupta (1997), Ringquist (1997), Brooks and Sethi (1997), Sadd et al. (1999), Baden et al. (2007), Shadbegian et al. (2007), and Baryshnikova (2010).

6. Studies also examine whether cleaning up a community increases housing prices. Households that the policy was originally intended to help may not be better off because they may be forced to either pay higher rent or migrate to other lower-cost neighborhoods with fewer amenities (e.g., Banzhaf and Walsh 2008; Depro and Timmins 2012; and Noonan 2012).

7. Evidence is mixed on whether regulatory enforcement varies with community socioeconomic characteristics. For instance, Gray et al. (2012) find little evidence that plants in poor and minority areas face different levels of inspection or enforcement. However, Sigman (2001) finds that a Superfund site is processed more quickly if it is located in a politically active or higher-income community, while Konisky (2009) finds that state enforcement of federal regulations varies with income but not by race. See chapter 7 for a more detailed review of this literature.

8. Vajjhala et al. (2008) notes the paucity of academic studies that evaluate proposed policies to address environmental justice. Most document whether environmental injustice currently exists.

9. An incremental effect is the change in the outcome of interest due to regulation relative to the baseline, which describes how emissions and other relevant factors would evolve without regulation.

10. While EO 12898 specifically mentions impacts on minority, low-income, and indigenous populations, it does not define these groups.

11. Children from households below the poverty threshold are more likely to receive care from parents or relatives, although it is unknown whether they will experience the benefits of reduced lead exposure from work practices mandated in child-occupied facilities.

12. PM2.5 refers to particulate matter (PM) smaller than 2.5 micrometers.

13. Fate and transport models are used to simulate or predict the distribution of concentrations in environmental media (e.g., air, water, and soil). These models range from a set of simple equations requiring little data to extremely complex sets of equations requiring detailed site-specific information.

14. The EPA used concentration-response functions that stratified impacts by education attainment because the epidemiological literature provided evidence that populations with less than grade 12 education have a higher risk of mortality than populations with higher levels of education.

References

Anderton, Douglas L., Andy B. Anderson, John Michael Oakes, and Michael R. Fraser. 1994. Environmental Equity: The Demographics of Dumping. *Demography* 31 (2): 229–248.

Arora, Seema, and Timothy N. Cason. 1998. Do Community Characteristics Influence Environmental Outcomes? Evidence from the Toxics Release Inventory. *Journal of Applied Econometrics* 1 (2): 413–453.

Asch, Peter, and Joseph J. Seneca. 1978. Some Evidence on the Distribution of Air Quality. *Land Economics* 54 (3): 278–297.

Ash, Michael, and T. Robert Fetter. 2004. Who Lives on the Wrong Side of the Environmental Tracks? Evidence from the EPA's Risk-Screening Environmental Indicators Tool. *Social Science Quarterly* 85 (2): 441–462.

Baden, Brett M., and Don L. Coursey. 2002. The Locality of Waste Sites within the City of Chicago: A Demographic, Social, and Economic Analysis. *Resource and Energy Economics* 24: 53–93.

Baden, Brett M., Douglas S. Noonan, and Rama Mohana R. Turaga. 2007. Scales of Justice: Is There a Geographic Bias in Environmental Equity Analysis? *Journal of Environmental Planning and Management* 50 (2): 163–185.

Banzhaf, H. Spencer. 2011. Regulatory Impact Analyses of Environmental Justice Effects. *Journal of Land Use & Environmental Law* 27 (1): 1–30.

Banzhaf, H. Spencer, and Randall P. Walsh. 2008. Do People Vote with Their Feet? An Empirical Test of Tiebout's Mechanism. *American Economic Review* 98 (3): 843–863.

Baryshnikova, Nadezhda V. 2010. Pollution Abatement and Environmental Equity: A Dynamic Study. *Journal of Urban Economics* 68 (2): 183–190.

Been, Vicki. 1994. Locally Undesirable Land Uses in Minority Neighborhoods: Disproportionate Siting or Market Dynamics? *Yale Law Journal* 103 (6): 1383–1421.

Been, Vicki, and Francis Gupta. 1997. Coming to the Nuisance or Going to the Barrios? A Longitudinal Analysis of Environmental Justice Claims. *Ecology Law Quarterly* 24 (1): 1–56.

Berry, Brian J. L. 1977. *Social Burdens of Environmental Pollution: A Comparative Metropolitan Data Source*. Cambridge, MA: Ballinger Publishing.

Bowen, William M. 2001. *Environmental Justice through Research-Based Decision-Making*. New York: Garland Publishing.

Bowen, William M., Mark J. Salling, Kingsley E. Haynes, and Ellen J. Cyran. 1995. Toward Environmental Justice: Spatial Equity in Ohio and Cleveland. *Annals of the Association of American Geographers* 85 (4): 641–663.

Brooks, Nancy, and Rajiv Sethi. 1997. The Distribution of Pollution: Community Characteristics and Exposure to Air Toxics. *Journal of Environmental Economics and Management* 32: 233–250.

Bullard, Robert D. 1983. Solid Waste Sites and the Black Houston Community. *Sociological Inquiry* 53:273–278.

California Environmental Protection Agency (EPA). 2010. Cumulative Impacts: Building a Scientific Foundation. December. http://oehha.ca.gov/ej/pdf/CIReport 123110.pdf.

Cameron, Trudy Ann, and Graham D. Crawford. 2003. Superfund Taint and Neighborhood Change: Ethnicity, Age Distributions, and Household Structure. *The Political Economy of Environmental Justice*, ed. H. Spencer Banzhaf, 137–169. Stanford, CA: Stanford University Press.

Chakraborty, Jayajit. 2011. Revisiting Tobler's First Law of Geography: Spatial Regression Models for Assessing Environmental Justice and Health Risk Disparities. *Geospatial Analysis of Environmental Health Geotechnologies and the Environment* 4: 337–356.

Chakraborty, Jayajit, and Juliana A. Maantay. 2011. Proximity Analysis for Exposure Assessment in Environmental Health Justice Research. In *Geospatial Analysis of Environmental Health*, ed. Juliana A. Maantay and Sara McLafferty, 111–138. Berlin: Springer-Verlag.

Chakraborty, Jayajit, Juliana A. Maantay, and Jean D. Brender. 2011. Disproportionate proximity to environmental health hazards: Methods, models, and measurement. *American Journal of Public Health* 101 (S1): S27–S36.

Cropper, Maureen L. 2010. "Incorporating Equity Concerns into Benefit Cost Analysis." Presentation. http://www.epa.gov/ncer/events/calendar/2010/mar17/presentations/cropper.pdf

Davidson, Pamela, and Douglas L. Anderton. 2000. Demographics of Dumping II: A National Environmental Equity Survey and the Distribution of Hazardous Materials Handlers. *Demography* 37 (4): 461–466.

Depro, Brooks, and Christopher. Timmins. 2012. Residential Mobility and Ozone Exposure: Challenges for Environmental Justice Policy. In *The Political Economy of Environmental Justice*, ed. H. Spencer Banzhaf, 115–136. Stanford: Stanford University Press.

Environmental Protection Agency (EPA). 2000. *Guidelines for Conducting Economic Analyses.* September. U.S. EPA, Office of the Administrator, Report #: EPA 240-R-00-003.

Environmental Protection Agency (EPA). 2004. *Toolkit for Assessing Potential Allegations of Environmental Injustice.* http://www.epa.gov/compliance/ej/resources/policy/ej-toolkit.pdf.

Environmental Protection Agency (EPA). 2010. *EPA's Action Development Process: Interim Guidance on Considering Environmental Justice during the*

Development of an Action. July. http://www.epa.gov/environmentaljustice/resources/policy/considering-ej-in-rulemaking-guide-07-2010.pdf.

Environmental Protection Agency (EPA). 2013a. *Plan EJ 2014 Progress Report.* February. http://www.epa.gov/environmentaljustice/resources/policy/plan-ej-2014/plan-ej-progress-report-2013.pdf.

Environmental Protection Agency (EPA). 2013b. *Technical Guidance for Assessing Environmental Justice in Regulatory Analysis.* May. http://www.epa.gov/environmentaljustice/plan-ej/rulemaking.html.

Environmental Protection Agency (EPA). 2014a. Environmental Justice, Children's Environmental Health, and Other Distributional Considerations; Chapter 10 in 2010 *Guidelines for Conducting Economic Analyses.* May. http://yosemite.epa.gov/ee/epa/eed.nsf/webpages/Guidelines.html.

Environmental Protection Agency (EPA). 2014b. *Plan EJ 2014 Progress Report.* February. http://www.epa.gov/compliance/environmentaljustice/resources/policy/plan-ej-2014/plan-ej-progress-report-2014.pdf.

Freeman, A. Myrick, III. 1969. Income Distribution and Social Choice: A Pragmatic Approach. *Public Choice* 7 (1): 3–21.

Freeman, A. Myrick, III. 1972. The Distribution of Environmental Quality. In *Environmental Quality Analysis: Theory and Method in the Social Sciences,* ed. Allen V. Kneese and Blair T. Bower, 243–280. Baltimore, MD: Johns Hopkins Univerity Press.

General Accounting Office (GAO). 1983. *Siting of Hazardous Waste Landfills and Their Correlation with Racial and Economic Status of Surrounding Communities.* Washington, DC: General Accounting Office (GAO).

General Accounting Office (GAO). 2005. *Environmental Justice: EPA Should Devote More Attention to Environmental Justice When Developing Clean Air Rules.* Report to the Ranking Member, Subcommittee on Environment and Hazardous Materials, Committee on Energy and Commerce, House of Representatives. GAO-05-289.

Gray, Wayne B., and Ronald J. Shadbegian. 2004. Optimal Pollution Abatement—Whose Benefits Matter and How Much? *Journal of Environmental Economics and Management* 47: 510–534.

Gray, Wayne B., and Ronald J. Shadbegian. 2007. The Environmental Performance of Polluting Plants: A Spatial Analysis. *Journal of Regional Science* 47: 63–84.

Gray, Wayne B., Ronald J. Shadbegian, and Ann Wolverton. 2012. Environmental Justice: Do Poor and Minority Populations Face More Hazards? In *The Oxford Handbook of the Economics of Poverty,* ed. Phillip N. Jefferson, 605–637. New York: Oxford University Press.

Hamilton, James T. 1995. Testing for Environmental Racism: Prejudice, Profits, Political Power? *Journal of Policy Analysis and Management* 14 (1): 107–132.

Harper, Sam, Eric Ruder, Henry A. Roman, Amelia Geggel, Onyemaechi Nweke, Devon Payne-Sturges, and Jonathan I. Levy. 2013. Using Inequality Measures to

Incorporate Environmental Justice into Regulatory Analyses. *International Journal of Environmental Research and Public Health* 10: 4039–4059.

Konisky, David M. 2009. Inequities in Enforcement? Environmental Justice and Government Performance. *Journal of Policy Analysis and Management* 28 (1): 102–121.

Konisky, David M., and Tyler S. Schario. 2010. Examining Environmental Justice in Facility-Level Regulatory Enforcement. *Social Science Quarterly* 91 (3): 835–855.

Kriesel, Warren, Terence J. Centner, and Andrew G. Keeler. 1996. Neighborhood Exposure to Toxic Releases: Are There Racial Inequalities? *Growth and Change* 27: 479–499.

Maguire, Kelly, and Glenn Sheriff. 2011. Comparing Distributions of Environmental Outcomes for Regulatory Environmental Justice Analysis. *International Journal of Environmental Research and Public Health* 8: 1707–1726.

Mohai, Paul, and Bunyan Bryant. 1992. Environmental Racism: Reviewing the Evidence. In *Race and the Incidence of Environmental Hazards*, ed. Bunyan Bryant and Paul Mohai. Boulder, CO: Westview Press.

Mohai, Paul, and Robin Saha. 2006. Reassessing Racial and Socioeconomic Disparities in Environmental Justice Research. *Demography* 43 (2): 383–399.

Morello-Frosch, Rachel, Manuel Pastor, and James Sadd. 2001. Environmental Justice and Southern California's "Riskscape": The Distribution of Air Toxics Exposures and Health Risks among Diverse Communities. *Urban Affairs Review* 36 (4): 551–578.

Noonan, Douglas S. 2012. Amenities Tomorrow: A Greenbelt Project's Impacts over Space and Time. In *The Political Economy of Environmental Justice*, ed. S. Banzhaf, 170–196. Stanford, CA: Stanford University Press.

Office of Management and Budget (OMB). 2003. *Circular A-4*. September. Washington, DC: White House. www.whitehouse.gov/omb/circulars_a004_a-4.

Ringquist, Evan J. 2005. Assessing Evidence of Environmental Inequities: A Meta-Analysis. *Journal of Policy Analysis and Management* 24 (2): 223–247.

Ringquist, Evan J. 1997. Equity and the Distribution of Environmental Risk: The Case of TRI Facilities. *Social Science Quarterly* 78 (4): 811–829.

Sadd, James L., Manuel Pastor, Jr., J. Thomas Boer, and Lori D. Snyder. 1999. "Every Breath You Take …" The Demographics of Toxic Air Releases in Southern California. *Economic Development Quarterly* 13 (2): 107–123.

Shadbegian, Ronald J., and Wayne B. Gray. 2012. Spatial Patterns in Regulatory Enforcement: Local Tests of Environmental Justice. In *The Political Economy of Environmental Justice*, ed. H. Spencer Banzhaf, 225–248. Stanford, CA: Stanford University Press.

Shadbegian, Ronald J., Wayne B. Gray, and Cynthia. Morgan. 2007. Benefits and Costs from Sulfur Dioxide Trading: A Distributional Analysis. In *Acid in the Environment: Lessons Learned and Future Prospects*, ed. G. Visgilio and D. Whitelaw. New York: Springer Press.

Sigman, Hilary. 2001. The Pace of Progress at Superfund Sites: Policy Goals and Interest Group Influence. *Journal of Law & Economics* 44: 315–344.

United Church of Christ (UCC). 1987. *Toxic Waste and Race in the United States.* UCC. New York: United Church of Christ Commission for Racial Justice.

Vajjhala, Shalini P., Amanda Van Epps, and Sarah Szambelan. 2008. Integrating EJ into Federal Policies and Programs: Examining the Role of Regulatory Impact Analyses and Environmental Impact Statements. RFF Discussion Paper, 08–45. Resources for the Future.

Wolverton, Ann. 2009. Effects of Socio-Economic and Input-Related Factors on Polluting Plants' Location Decisions. *Berkeley Electronic Journal of Economic Analysis and Policy, Advances* 1. http://www.degruyter.com/view/j/bejeap.2009.9.1/bejeap.2009.9.1.2083/bejeap.2009.9.1.2083.xml.

Wolverton, Ann. 2012. The Role of Demographics and Cost Related Factors in Determining Where Plants Locate: A Tale of Two Texas Cities. In *The Political Economy of Environmental Justice*, ed. S. Banzhaf. Stanford, CA: Stanford University Press.

Zimmerman, Rae. 1993. Social Equity and Environmental Risk. *Risk Analysis* 13 (6): 649–666.

6

Public Participation and Environmental Justice: Access to Federal Decision Making

Dorothy M. Daley and Tony G. Reames

Executive Order 12898 on Environmental Justice (EO 12898) directs federal agencies to develop broad strategies to identify and address any disproportionate, negative impacts stemming from agency activities and affecting minority and low-income communities. It aims to integrate environmental justice considerations into the standard operating procedures of federal agencies, and in doing so, mitigate the likelihood that minority and low-income communities experience concentrated environmental burdens. By their very nature, environmental justice concerns are not uniformly distributed across the country. Rather, they tend to be local or regional in scale, and the specific details of the environmental justice situation may be unique. For example, contaminated drinking water in the Appalachians, poor air quality in Southern California, or traffic congestion in Houston may disproportionately affect low-income communities, minority communities, or both, but these problems occur on a local or regional scale. Therefore, a "one size fits all" approach to environmental justice is unlikely to be efficient or effective. The diversity and distribution of environmental justice problems pose a considerable challenge to federal agencies charged with implementing the executive order.

Public participation is one mechanism to address this challenge. Effective public participation could not only help identify and characterize environmental justice concerns, but it could also inject critical local knowledge and inform policies and programs to solve environmental justice problems (Dietz et al. 2008). EO 12898 emphasizes the need for widespread public participation to better identify, understand, and tackle environmental justice concerns. Consequently, the definition of environmental justice by the Environmental Protection Agency (EPA) underscores the importance of public participation by calling for "meaningful involvement of all people" in every aspect of environmental decision making.

At the federal level, cultivating and maintaining widespread and diverse public participation to prevent and combat concentrated environmental risk requires significant agency commitment (Dietz at al 2008; Innes and Booher 2004; Foreman 1998). Although EO 12898 tasks federal agencies to increase public participation, to date, there has been limited evaluation aimed at understanding how—after twenty years of implementation— agencies have addressed that charge. This chapter evaluates the ways in which federal agencies utilize public participation to address environmental justice in minority and low-income communities in response to EO 12898.

The chapter begins by examining the research on public participation and environmental decision making in general. Based on this, we highlight the opportunities and challenges of using participation to address environmental justice problems. To better understand how EO 12898 has influenced public participation in federal agencies, we examine the EPA, the agency responsible for leading the federal government's approach on environmental justice. In addition, the chapter compares and contrasts how the EPA, the Department of Energy (DOE), and the Department of Transportation (DOT) utilize public participation to advance environmental justice. Our assessment of these agencies is based upon examining a range of government documents and electronic sources that detail each agency's environmental justice strategy in 1995 and subsequent revisions. In addition, we compiled annual reports from the three agencies, along with any existing evaluations of agency efforts related to implementing the executive order. Material from these sources is integrated with findings from existing research on civic engagement, public participation, and social equity to provide a general assessment of how well each agency has expanded opportunities for public participation in light of the executive order. Finally, the chapter concludes with policy prescriptions for federal agencies as they move forward with their reaffirmed commitment to environmental justice and public participation.

Public Participation and Environmental Decision Making

Public participation is woven into the fabric of American democracy and it can take a variety of forms. Voting, letter writing, attending public meetings, forming an interest group, or serving on advisory committees are all examples of public participation. In recent decades, public participation has been an increasingly important element of environmental governance (Bulkeley and Mol 2003; Daley 2013; Dietz et al. 2008). Public participation in environmental decision making can be *process-oriented*. For example, public comment periods for permitting, setting standards, and writing and revising regulations are all avenues to engage the public in the process of environmental decision making. In addition, when the EPA or other agencies are in the midst of a particularly controversial environmental issue, they may hold listening sessions, workshops, and seminars to provide additional opportunities to gauge public sentiment on potential changes in advance of a final decision. Public participation can also be more *outcome-oriented*. This type of participation tends to focus on existing environmental hazards resulting from past decisions. Examples include protesting the existence of multiple polluting facilities within one community or serving on a community advisory board for a hazardous waste site that needs to be remediated. Outcome-oriented participation can also advocate for strict enforcement and compliance with existing environmental laws by reporting facilities that might be in violation of those laws.

Environmental decisions tend to be highly technical, complex, and value-laden. Scientific and technical expertise is critical to foster better understanding of the nature of environmental problems and potential solutions. But even when scientific and technical knowledge is combined to inform environmental decisions, conflict and gridlock can remain, particularly when public preferences on the nature and distribution of environmental risks and benefits are poorly understood or not carefully considered (Dietz et al. 2008; Weber 2003; Webler and Tuler 2006). Public participation provides an avenue to gauge public preferences and insert public values into environmental decision making. Recent research suggests that when done well—an important caveat—public participation can increase equity, reduce conflict and gridlock, and lead to improved environmental decision making (Dietz et al. 2008; Klyza and Sousa 2013; Pellow and Brulle 2005). There are a number of ways that public participation could result in these desirable outcomes. Widespread and meaningful participation can add legitimacy to any final decision and enhance overall levels of trust in government. Local knowledge could provide more detailed information on the nature of a problem, along with identifying locally appropriate solutions that capitalize on community norms and in doing so, increase the chances of successful policy implementation (Bulkeley and Mol 2003; Daley 2013; Dietz et al. 2008; Reed 2008).

Conversely, when done poorly, public participation can reduce overall levels of trust in government, increase conflict and gridlock, and highlight weaknesses in both the decision-making process and outcomes from that process. Thus, the promise of public participation hinges on both the quality of the participatory process and the overall ability of stakeholders to engage in the process. Effective or high-quality participatory processes tend to have a clear goal, adequate financial and human capital, consistent institutional commitment, an ability for participatory action to influence all stages of decision making, and an emphasis on implementation and evaluation (Beierle and Cayford 2002; Charnley and Engelbert 2005; Dietz et al. 2008; Leach 2006; Reed 2008; and Sirianni 2009).

Most environmental legislation calls for some form of public participation, and EO 12898 is no exception. Traditional participatory mechanisms used in environmental legislation include public hearings and public comment periods for permits, regulations, and other environmental decisions. One-time seminars and workshops, occasional roundtables, and advisory committees are increasingly being utilized to improve the scope and impact of public participation (Innes and Booher 2004; Kellogg and Mathur 2003; Spyke 1999; Beierle 1998; Coglianese 1997). Although some of these approaches to public participation can be more innovative than others, research suggests that as currently implemented, all of these mechanisms tend to result in top-down, expert-driven decision making, despite consistent calls for more bottom-up approaches to public participation (Agyeman and Angus 2003; Dietz et al. 2008; Laurian 2007). Therefore, in their present form, these participatory mechanisms would not be *effective* participatory processes to reduce conflict and gridlock or improve equity (Beierle 1998; Brion 1988; Buchanan 2010; Fine and Owen 2005; Hernandez 1995; Webler and Tuler 2006).

In many instances, both the quality of the participatory process and the ability of stakeholders to engage in that process are compromised (Klyza and Sousa 2013). Determining who represents the affected public and how to broaden the scope of public participants in a decision-making process remains difficult. Government agencies commonly recruit people from organized groups who have expressed interest or are active around a particular issue (Larson and Lach 2008). Although this seems like an efficient strategy, it raises concern about the degree to which organized interests represent the views of a broader community. Relying on this approach to insert public values into decision making may not expand access to new participants, as it tends to result in a similar set of engaged stakeholders taking part in the decision-making process (Coglianise 2006; Agyeman and Angus 2003; Beierle 1998). Comparatively, business interests are much easier to identify and engage in the decision-making process. They have considerable incentive to remain involved in environmental decision making, particularly when decisions directly regulate their behavior. Moreover, these concentrated interests tend to have more resources than other stakeholders, and they are in a better position to shape environmental decision making in their favor (Furlong 2007; Kamieniecki 2006; Yackee and Yackee 2006).

Federal agencies face a complicated decision-making environment: While regulated interests are motivated and likely to remain active in any and all aspects of decision making, engaging the public is far more challenging. Overall levels of civic engagement and public participation in the United States are declining (Macedo et al. 2005). Inconsistent public participation directly limits the ability of this tool to give voice to a diverse public (Dietz et al. 2008; Foreman 1998). Meaningful public participation within federal agencies requires creativity, resources, and a consistent commitment to a participatory process. Since the Reagan era, the federal government has experienced near-constant pressure to devolve responsibilities to state and local government whenever possible. While devolution to subnational governments may provide more opportunity for public participation in state and local institutions (Macedo et al. 2005), it significantly complicates the ability of federal agencies to maintain well-resourced public participation processes.

Public Participation and Environmental Justice: Opportunities and Challenges

Successfully incorporating public participation into any environmental decision-making process remains a significant opportunity and a challenge. These opportunities and challenges are magnified with environmental justice concerns. Because public participation can increase equity and reduce conflict and gridlock in decision making, it holds tremendous promise for both ameliorating existing environmental justice problems and preventing future concentrated risks. However, this promise is situated in a broader context of racial, economic, and political inequality (Brulle and Pellow 2006; Cole and Foster 2001; Schlozman, Verba, and Brady 2012). Minority and low-income communities are less politically active overall (Rosenstone and Hansen 1993; Schlozman, Verba, and Brady 2012), and historically, they have mobilized less around environmental issues than wealthy, white communities have (Brulle and Pellow 2006; Gauna 1995). The modern environmental movement is largely comprised of white, middle class Americans concerned with natural resource degradation (Baber and Bartlett 2013; Gauna 1995); often, social justice and equity are not on the agenda (Mohai, Pellow, and Timmons Roberts 2009). In contrast, the environmental justice movement is distinctly shaped by the history of civil rights in the United States, as well as grassroots anti-toxics activism stemming from environmental disasters like Love Canal. As a result, there tends to be more outcome-oriented public participation in minority and low-income communities that focus on existing environmental risks in their communities.

Some within the environmental justice community view the disproportionate impact of environmental hazards on minority and low-income communities as a product of the same social structure that produces racial oppression. Others view the concentration of burdens as an intended consequence of dominant economic forces (Brulle and Pellow 2006; Cole and Foster 2001; Foreman 1998). Regardless, either of these causal stories creates significant hurdles in using public participation to advance equity in environmental decision making. Successful public participation in environmental decision making is predicated on trust. As with any collaborative relationship, high levels of trust facilitate productive interactions (Lubell et al. 2005). When that trust is frayed, the promise that public participation would increase equity and reduce conflict and gridlock in decision making becomes more difficult to realize.

Despite this difficulty, public participation can provide critical information regarding the ways in which local residents experience concentrated environmental burdens. From a pragmatic point of view, the expansive geography of the United States and the local or regional scale of environmental justice concerns combine to highlight the need for public participation as an important mechanism to advance environmental justice in federal agencies. But patterns of public participation directly affect the ability of this tool to ameliorate environmental justice concerns. In fact, some contend that public participation in a decentralized political system actually contributes to environmental justice problems (Foreman 1998; Gauna 1995; Munton 1996).

Local grassroots mobilization, described as "Not In My Backyard" or NIMBY, has been a powerful response to environmental risks (Fletcher 2003; Kraft and Clary 1991). Over the years, politically active communities have mobilized successfully to block power plants, hazardous waste sites, landfills, and other locally unwanted land uses (LULUs) from being established in their vicinity. This participatory activity increases the likelihood that LULUs are instead located in communities with lower levels of political mobilization and public participation, which often are minority and low-income neighborhoods. NIMBY can extend beyond siting; local groups may advocate stringent enforcement of laws governing existing facilities within their communities and demand swift action in response to accidents (Gauna 1995), while less organized communities receive less attention. Thus, this type of grassroots activism can facilitate the concentration of environmental risks in minority and low-income communities in multiple ways.

The past several decades have witnessed a consistent increase in opportunities for public participation in almost all aspects of environmental decision making. However, there is little evidence of a corresponding increase in widespread, diverse public involvement. In fact, evidence suggests that levels of civic engagement and public involvement in the United States have declined (Coglianese 2006; Putnam 2000; Sander and Putnam 2010), and existing patterns of participation tend to represent white, wealthy, and educated citizens (Macedo et al. 2005; Schlozman, Verba, and Brady 2012). Minority and low-income communities may face higher barriers to entry for public participation. Public participation requires some general understanding of how agency decisions affect citizens' interests. Lower levels of educational attainment, language barriers (Larson and Lach 2008; Fine and Owen 2005), and limited knowledge about federal decision-making processes or how decisions affect their lives are likely to constrain citizens' motivation to participate (Coglianese 2006).

Despite these challenges, public participation remains an important tool in advancing environmental justice considerations within federal agencies. In 2011, the seventeen agencies covered under EO 12898 declared, "the continued importance of identifying and addressing environmental justice considerations in agency programs, policies and activities"[1] by signing a Memorandum of Understanding on Environmental Justice and Executive Order 12898.

The EPA, Public Participation, and Environmental Justice

The EPA is the one of the most active federal agencies in terms of incorporating environmental justice into agency activities, and it is also responsible for convening and leading the Interagency Working Group on Environmental Justice. Despite a concerted effort to integrate environmental justice and public participation into agency activities, progress remains uneven (GAO 2011; CCR 2003). In part, this reflects the nature of the participation challenges outlined above, as well as highlighting the challenge of changing bureaucratic behavior. The EPA is a large bureaucracy charged with implementing a wide variety of environmental legislation, including the Clean Water Act (CWA), the Clean Air Act (CAA), the Resource Conservation and Recovery Act (RCRA), the Toxic Substance Control Act (TSCA), and several others, in addition to EO 12898. The agency's duties are scientifically complex and administratively difficult, with decision making often generating conflict across a range of interests, including environmental justice stakeholders. Environmental justice issues may arise in a number of activities for which the agency is responsible, including setting standards, permitting facilities, awarding grants, issuing licenses and regulations, and reviewing proposed actions by other federal agencies.

Public participation is a key element underpinning the EPA's environmental justice strategy as the agency seeks to "work with communities through communication, partnership, research and the public participation process" and also "help affected communities have access to information which will enable them to meaningfully participate in activities" (EPA 1995, p. 3). Shortly after EO 12898 was signed, the agency acknowledged that a host of stakeholders are directly affected by the agency's environmental decisions and therefore "must have every opportunity for public participation in the making of those decisions" (EPA 1995, p. 4). Environmental justice problems tend to be local in nature; therefore, participation and outreach efforts need to be tailored to the specific problem and community context. But this type of customization is complicated. In the first ten years following EO 12898, EPA regional offices relied on different approaches to identifying minority and low-income communities; this variation created significant implementation and evaluation hurdles (EPA 2004). As research on public participation suggests and the EPA's experience highlights, identifying the affected community is challenging (Daley 2013; Dietz et al. 2008; Larson and Lach 2008), and absent this identification, public participation is not likely to yield increases in equity.

Environmental justice staff within the agency have a challenging mission: to integrate environmental justice considerations into all aspects of a major bureaucratic organization and create systemic change, including increasing opportunities for meaningful public participation. Historically, the EPA has been a "stove-piped" organization with separate offices for air, water, and toxics, each operating largely in their own silos. The agency has no overarching legislative agenda, but rather a series of diverse environmental problems to address. Without clear guidance on how to prioritize among divergent environmental efforts, the agency has been criticized as engaging in turf battles (Rosenbaum 2002). Creating meaningful change in a large hierarchical organization requires significant time (up to a decade or longer), consistent commitment from all levels of personnel, considerable resources, and multiple, overlapping strategies (Bardach 1998; Brehm and Gates 2002; Mazmanian and Sabatier 1989; Wilson 1989).

In November 1992, an Office of Environmental Equity (later renamed the Office of Environmental Justice (OEJ)) was created within the agency, and each regional office has environmental justice coordinators to integrate environmental justice into its policies, programs, and activities. Over the last two decades, the agency has relied upon a range of participatory mechanisms to engage minority and low-income communities in environmental decision making. These include public meetings, public notice and comment periods, developing partnerships, conducting seminars and workshops, and convening the National Environmental Justice Advisory Council (NEJAC). Since 1993, NEJAC has been providing independent advice and recommendations to the agency on environmental justice issues. The approximately 26 members of the committee represent stakeholders from academia, business and industry, state and local governments, tribal governments, environmental organizations, community groups, and nongovernmental organizations.

Agency and NEJAC documents note the importance of public participation, and several also highlight successful collaborative relationships, but it is difficult to determine if these success stories are representative of the norm (EPA 1997). Similarly, it is difficult to determine if EPA's general approach to participation has consistently resulted in increased diversity of participation. In 2006, the agency released a series of environmental justice accomplishment reports on their website to highlight environmental justice integration, including public participation, into the agency's decision-making processes.[2] Notably, two (out of ten) regional offices and three (out of nine) offices within the agency's headquarters do not provide accomplishment reports. The available reports provide detailed tables listing the region's or office's activity, output, outcome, and results in relationship to a set of goals and objectives. They provide an excruciating level of detail, but this actually makes understanding the overall progress toward integration more difficult.

For example, one objective from the Office of Air and Radiation (OAR) report included providing opportunities for meaningful involvement between agency staff and affected community members. OAR listed six activities they developed to achieve this goal, one of which was the development of training modules to facilitate public participation in air quality permitting in environmental justice communities. Another activity included year-round monitoring and reporting of the Air Quality Index (AQI). The training module to enhance engagement in air quality permitting was not funded and therefore had no impact on participation. But year-round monitoring and reporting of the AQI led to the development of an e-alert system. This allows individuals to receive electronic air quality information, which can be particularly useful for limiting exposure to air pollution. The report does not indicate if the agency engaged environmental justice communities to encourage participation in the e-alert program. They may have done so, and it is simply not listed in this report. Absent more context about which activities are critical in achieving meaningful involvement, it is hard to evaluate the accomplishments listed in these reports.

Along with using traditional participatory approaches such as public notice and comment periods, the agency has invested considerable effort in capacity building by improving access to environmental and demographic information. Providing information to affected communities and interested citizens is one mechanism to level the playing field and facilitate effective participation. EJView, for example, is an interactive mapping tool on the EPA's website (http://epamap14.epa.gov/ejmap/entry.html) which allows users to see how environmental burdens may be concentrated within a geographic area. Any facility reporting to EPA can be identified on a map, along with water-monitoring stations. This includes current hazardous waste transfer, storage and disposal facilities, Superfund sites, Brownfield areas, Toxics Release Inventory (TRI) facilities, and air and water dischargers. This tool overlays demographic information as well, making it relatively easy to identify if low-income or minority communities have a large number of EPA-permitted facilities compared to wealthy, white communities. EJView also provides critical health outcomes, such as information on infant mortality rates, birth rates, and cancer risk. This is a powerful diagnostic tool that communities and advocacy groups could use to identify potential environmental justice issues. In its current form, however, it falls short. While the agency has devoted considerable resources to create this level of access to information, EJView lacks any corresponding information about how to *act* on this information. This is a missed opportunity for the agency to facilitate participation. If people use EJView to better understand the density of permitted facilities within their community, what next steps could they take to engage in environmental decision making? Currently, there is no guidance connecting the information provided in EJView to information on the agency's permitting process, rulemaking, or other elements of environmental decision making.

In comparison, the EPA's Community Action for a Renewed Environment (CARE) program exemplifies a model initiative to provide information and empower communities while also facilitating public engagement (Hansell, Hollander, and John 2009; Sirianni 2009; EPA 2011). This competitive grant program aims to reduce toxic pollution by (1) working collaboratively with local communities to reduce exposure to toxic pollutants, (2) helping communities understand individual and cumulative sources of toxic exposure, (3) working with communities to identify and prioritize risk-reducing activities, and (4) creating long-term, community-based partnerships to improve and protect the local environment.[3] While many federal agencies, including the EPA, struggle with how best to work with local communities, CARE demonstrates that national-local partnerships can be effective and have considerable spillover benefits. For example, two different CARE communities identified exposure to solvents and other chemicals used in auto body shops as an environmental risk that could—and should—be minimized. On the local level, environmental groups, the EPA, and local auto body shops worked collaboratively to identify alternative, less toxic chemicals for paint stripping, along with exploring opportunities for improved disposal practices. This experience, in turn, prompted the EPA to work closely with these CARE grantees to develop national emission standards regulating paint strippers and other solvents used in auto body refinishing (Hansell, Hollander, and John 2009). These regulations broaden the impact of the CARE program because they apply to auto body shops across the country.

The focus of the CARE program often results in investment in environmental justice communities, but any community can apply for support from this program. The Environmental Justice Small Grants program, administered by the EPA since 1994, is specifically designed to support environmental justice communities. These grants provide a tangible way to identify local issues, support community groups, and build trust between the agency and affected community members. Reflecting the diversity of environmental justice concerns, grants have focused on a wide range of environmental issues, such as radon, lead, farmworker safety, recycling, water quality, and children's health.[4] But the commitment to investing in communities has varied greatly over time. Figure 6.1 shows the number of Environmental Justice Small Grants awarded by year. The number of grants peaked shortly after EO 12898 was signed, and while there were increases around the time of the American Recovery and Reinvestment Act, more recent years have witnessed a decline. It is

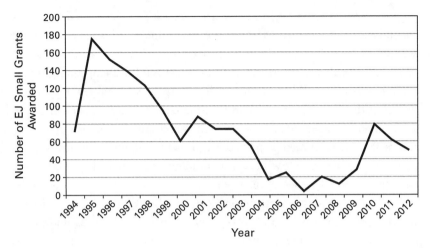

Figure 6.1

Number of Environmental Justice (EJ) Small Grants Awarded by Year (1994–2012).

Data in this chart are drawn from reports available at http://www.epa.gov/environmentaljustice/grants/ej-smgrants.html.

not possible to conclude if the variability in grants awarded reflects drastic changes in the applicant pool, changes in agency priorities, or is simply the result of federal budgetary constraints.

Moreover, recent research comparing all Environmental Justice Small Grants awarded between 1994 and 2004 suggest some challenges in targeting environmental justice communities. During this time period, the majority of the funds distributed were not directed toward minority and low-income communities with higher than the nation's average TRI emissions (Vajjhala 2010). It is impossible to determine if this pattern is due to a selection effect, meaning that those minority and low-income communities with above-average TRI emissions did not apply for a grant, or if in fact, there are problematic patterns in the distribution of funds. The Environmental Justice Small Grants program could be an innovative tool that builds trust and increases participation, but inconsistent support or biased implementation will limit its reach; effective participatory mechanisms require consistent commitment.

Environmental justice problems are diverse and may require that the EPA devote a range of expertise within its agency to understand and address community concerns. Agency leadership and commitment to high-quality participatory practices are necessary to tackle most environmental justice problems. Many in the environmental justice community worried that in 2001, under President George W. Bush, that EO 12898 would be repealed. This, in fact, did not occur. But the change in presidential administration did result in a shift in emphasis within environmental justice at the EPA. Christine Todd Whitman, the first administrator of the EPA under President Bush, issued a memo restating the agency's commitment to environmental justice. Rather than highlighting the significance of minority and low-income communities in environmental justice, the agency opted to focus on environmental justice for everyone (EPA 2004; Mohai, Pellow, and Timmons Roberts 2009). This change in focus, deemphasizing minority and low-income communities, adds additional challenges to expanding meaningful access for public participation within agency decision making. It strains trust in public institutions with stakeholders who were likely already suspicious of the federal government's commitment to social equity and environmental justice.

Under President Barack Obama, the EPA has renewed its focus on environmental justice and reiterated the importance of engaging minority and low-income communities to achieve environmental justice. In 2010, EPA administrator Lisa Jackson listed environmental justice as one of the agency's top priorities, and the following year, the agency published its blueprint to achieve environmental justice, *Plan EJ 2014*. This plan is an innovative strategy, and its structure suggests that agency leadership have clearly received the message: agency fragmentation is a significant hurdle in advancing environmental justice. *Plan EJ 2014* emphasizes cross-agency responsibility for advancing environmental justice into all aspects of environmental decision making. The plan also aims to improve the scientific and technical tools available to better diagnose and prevent the concentration of environmental risk in any community, along with evaluating more carefully the way that the agency's current initiatives can be tailored to support environmental justice goals.

Public participation looms large in this plan; the three main goals include to "protect the environment and health in overburdened communities; empower communities to take action to improve their health and environment; and establish partnerships with local, state, tribal, and federal governments and organizations to achieve healthy and sustainable communities." (EPA 2011, p. 1). *Plan EJ 2014* highlights the need to strengthen community-based programs with a particular focus on identifying "scalable and replicable" programs to more effectively address concentrated environmental burdens. For example, successful programs like CARE should be replicated and expanded whenever possible, and as the agency proceeds with implementing *Plan EJ 2014*, it intends to evaluate its progress regularly, along with producing information on "lessons learned" to facilitate communication about strategies that are effective in addressing environmental justice problems.

Some of the more innovative participatory approaches have emerged in recent years as the agency reinvigorates its commitment to environmental justice. In advance of developing *Plan EJ 2014*, the agency held a series of community forums and listening sessions across the country. It has also integrated communication technology to expand public participation, holding quarterly community outreach teleconference calls to gather and respond to community concerns. The OEJ hosts these calls, and agency staff from other programs and offices are present to respond to concerns that were submitted beforehand, as well as answer questions from the public during the call. Once a call is completed, the EPA provides both an audio version and a written transcript of it on the agency's website. Putting aside any consideration of the digital divide, this type of community outreach has the ability to expand access to a wide range of interested citizens. Recently available transcripts suggest that fragmentation within the agency and between the EPA and its state counterparts remains frustrating for community members trying to engage in environmental decision making. One caller noted that existing rules from the Superfund program, combined with longstanding agreements between the EPA and a state environmental agency, were contributing to the concentration of drinking water risk in a community, rather than protecting community members and providing a platform for a safe environment.[5]

While a renewed commitment to environmental justice is clear in agency documents, advancing a streamlined collaborative approach in a large organization remains difficult. *Plan EJ 2014* holds tremendous promise for advancing environmental justice. With an emphasis on cross-cutting implementation and improved access to information, *Plan EJ 2014* could enhance the diversity of public participation in environmental decision making. It aims to strike a balance between creating some standard operating procedures—like a national approach to identifying environmental justice communities—and providing discretion and flexibility to ensure that solutions are tailored to local problems.

As with all innovative plans promising change, the devil is the implementation details (Mazmanian and Sabatier 1989; Pressman and Wildavsky 1984). In the coming years, the challenge within EPA includes maintaining this renewed focus on environmental justice. Working across the breadth of programs within the agency and with stakeholders outside the agency to more successfully integrate public participation and environmental justice into the agency's decision-making processes will require resources and dedication. And, although the agency has had some success in expanding the conversation, improving access to information, and forging new partnerships with communities, more work remains, as the scope of environmental justice problems is large and public participation provides one potential mechanism for successful resolution of conflict and improved decision making.

Federal Agencies, Public Participation and Environmental Justice

While the EPA is a lead agency in advancing environmental justice within the federal government, it is by no means the only federal agency that struggles to integrate EO 12898 (including its focus on public participation) into daily routines. To compare and contrast the approach of other federal agencies to the EPA, we examine environmental justice and public participation in two other agencies: the DOE and the DOT. Table 6.1 compares the original participatory goals for the three agencies immediately after EO 12898 was signed, along with current participation goals. Material for this table is drawn exclusively from the government documents cited in the text. The table also highlights some of the participatory approaches commonly used by the agencies to address environmental concerns.

The Department of Energy
The DOE's mission is to "ensure America's security and prosperity by addressing its energy, environmental, and nuclear challenges through transformative science and technology solutions."[6] DOE has approximately 16,000 employees based around the country at its various offices, laboratories, and field sites. Among other duties, DOE is responsible for

Table 6.1

Comparison of Environmental Justice and Public Participation across Federal Agencies

Agency	EPA	DOE	DOT
Agency mission	Protect human health and the environment.	Ensure the nation's security and prosperity using science and technology to address energy, environmental, and nuclear challenges.	Ensure fast, safe, efficient, accessible, and convenient transportation systems.
Public participation goal (1995 EJ strategy)	To achieve widespread opportunity for participation in environmental decision making. An informed and involved community must underpin environmental protection.	Improve DOE's credibility and trust by enhancing public participation in agency activities.	Bring government decisions closer to the communities affected by them; expand opportunities for public participation in decisions relating to human health and the environment.
Public participation goal (updated strategy)	Public participation underpins two of the three main goals in *Plan EJ 2014*.	In more recent documents, the DOE's public participation goals remain unchanged.	To ensure the full and fair participation by all potentially affected communities in the transportation decision-making process. (DOT 2012 EJ strategy)
Participatory tools used	• Notice and comment periods • Advisory boards (NEJAC and community advisory boards) • EJ Small Grants • Translated documents and interpreter • Hot lines/reporting links • Conference calls	• Notice and comment periods • Site-specific advisory boards • Partnerships with minority organizations • Cooperative agreements to provide funding to communities	• Notice and comment periods • Seminars/workshops • Directs states, MPOs, and transit operators to engage citizens in project planning

cleaning up the environmental legacy from the nation's nuclear weapons program. This is an extraordinarily technically complex program that must confront very long time horizons, as the half-life on some of the spent nuclear material will remain radioactive for many decades to come. DOE is responsible for more than 80 federal facilities throughout the country that are contaminated and are currently undergoing environmental cleanup. In addition to developing its own environmental justice strategy, EO 12898 tasks the Secretary of Energy or a designated officer to serve on the Interagency Working Group.

Interestingly, DOE's public participation goal is to "enhance the credibility and public trust of the department by making public participation a fundamental component of all program operations, planning activities, and decision making" (DOE 1995). While the strategy emphasizes community participation and empowerment, it seems to suggest that this is done to improve the agency's profile, as opposed to improving or resolving conflict between the department and its stakeholders. Indeed, when the department first published its environmental justice strategy in 1995 following EO 12898, it noted that no stakeholder comments were included in the document because the department did not provide adequate time for public review prior to publishing the strategy.[7] Certainly, this is not an ideal way to establish positive relationships with stakeholders (Dietz et al. 2008). To its credit, DOE noted in early documents that the department's environmental justice strategy is a living document, and there would be ample opportunity for stakeholder contribution and change in the future.

Improving trust and credibility through participation may be the part of the department's approach to environmental justice, but it has not been easy. DOE acknowledges that minority, low-income, and tribal communities have traditionally lacked access to information and technical advisers to be informed participants in complex environmental decisions. NEJAC released a report noting, among other things, the strained relationships between DOE and the affected communities near the federal facilities the department manages (NEJAC 2004). National security and confidentiality issues surrounding activities at DOE facilities directly hamper the ability of the department to build trust and engage stakeholders. Despite this challenge, NEJAC recommended continued investment in public participation.

Currently, the Office of Legacy Management within DOE provides leadership for environmental justice programming. Environmental justice concerns may arise in relation to the treatment, storage, and disposal of hazardous, radioactive, and mixed waste. DOE is responsible for a number of federal sites throughout the nation that are currently undergoing cleanup processes, such as the Hanford site in Washington and the Savannah River site in South Carolina. Environmental justice issues may also arise during the National Environmental Policy Act (NEPA) planning process when the agency is upgrading existing energy projects or constructing new facilities.

DOE utilizes typical participatory mechanisms to engage the public, including public notice and comment periods, along with relying on citizen advisory groups and site-specific advisory boards. The department attempts to ensure that advisory boards reflect the communities they represent, but identifying and engaging an affected community can be difficult. For example, the Nevada Test Site Advisory Board has lacked broad representation for decades. The department attempted to recruit a wide cross section of potential stakeholders, but has had little success. In 2010, the Nevada Site office conducted a membership recruitment drive, advertising in the Las Vegas valley, rural communities, and a Spanish-language newspaper; however, no applications for the advisory board were received (DOE 2011). On the other hand, the Hanford Advisory Board reflects the diverse viewpoints in the affected community and region, including minorities and members of tribal nations (DOE 2011).

While the EPA relies on small grants to invest broadly in communities and environmental justice solutions, DOE takes a different programmatic approach. Many of their public participation initiatives seem to hinge on building technical capacity in affected communities (DOE 2011). This likely reflects both the department's mission to rely on "science and technology" to meet energy challenges, along with the highly technical nature of the cleanup at DOE facilities. The department supports efforts by historically black colleges and universities to build the science, technology, and engineering workforce and fills a variety of intern positions with minority students. In 2009, DOE entered into fifteen cooperative agreements with tribal nations, totaling approximately $6 million. This financial support is designed to build local communities' capacity to increase participation in DOE decision making (DOE 2011). Communities can use the funds to hire scientific and technical staff to help examine site cleanup strategies in order to educate community members on the more technical nature of DOE's responsibilities, as well as building trust by increasing access and transparency to site documents and the decision-making process.

In some cases, tribal nations have succeeded in affecting the department's decision making through cooperative agreements. For example, at the Hanford site, the inclusion of tribal input resulted in the protection of cultural, religious, and natural resources relating to Gable Mountain (DOE 2009). DOE's Los Alamos Pueblos Project in New Mexico relies on tribal governments to manage pollution monitoring programs and tribal governments also actively participate in generating the Los Alamos and National Laboratory Site-Wide Environmental Impact Statement by reviewing and commenting on the document (EPA 2013; DOE 2008). Despite success in many areas relating to environmental justice, DOE progress reports note that much work remains (DOE 2009; 2011).

The Department of Transportation

The DOT's mission is to "serve the United States by ensuring a fast, safe, efficient, accessible and convenient transportation system that meets our vital national interests and enhances the quality life of the American people, today and into the future.[8]" DOT is the largest of the three agencies compared here, with over 57,000 employees across the country. Within DOT, there are thirteen agencies, ranging from the Federal Highway Administration to the Federal Transit Administration to the Pipeline and Hazardous Material Safety Administration, working to accomplish DOT's broad mission. In addition to developing its own environmental justice strategy, EO 12898 tasks the Secretary of Transportation or a designated officer to serve on the Interagency Working Group.

In its 1995 environmental justice strategy, DOT expressed its commitment to "bringing government decision making closer to the communities and people affected by [its] decisions and ensuring opportunities for greater public participation in decisions relating to human health and the environment" (DOT 1995). DOT published its draft strategy in the Federal Register and mailed approximately 3,000 copies to constituent groups and representatives of the environmental justice community. They received approximately 50 comments and modified the final version to incorporate suggestions. In 1997, DOT issued an internal order on environmental justice as key component of the department's strategy.[9] In the internal order, the department reaffirmed its commitment to establishing or expanding procedures that enhance the participation of minority and low-income communities during all stages of the department's decision making. DOT updated its strategy in 2012, stating that its guiding principle on public participation is "to ensure the full and fair participation by all potentially affected communities in the transportation decision-making process" (DOT 2012).

Most of DOT's work involves setting policies and procedures to be followed by state and local governments in planning and constructing transportation facilities that receive federal funding. Stakeholder involvement, as required by DOT guidelines, is often carried out by state and local agencies responsible for the process. Environmental justice issues arise most frequently when some communities get transportation benefits, while others experience fewer, or when some communities suffer disproportionate negative impacts from transportation programs. DOT, like the EPA and DOE, struggles with representation in public participation, as some communities are consistently less well represented than others during policy and decision making concerning transportation resources (Cairns et al. 2003).

In the late 1990s, DOT acknowledged increasing public concerns regarding compliance with environmental justice provisions during metropolitan and statewide transportation planning (DOT 1999). In response, the department requested that administrators within DOT raise a litany of questions during state, metropolitan planning organization (MPO), and transit operator certifications pertaining to equity and public involvement in particular. Recognizing room for improvement in engaging low-income and minority communities in the public participation process, the department directed administrators to review public participation plans for agencies receiving federal funds and to address the strengths and deficiencies (DOT 1999).

In recent years, DOT has released "Environmental Justice Implementation Reports" available on the agency's website.[10] Like EPA and DOE, DOT engages in the typical range of public participation processes, including public notice and comment periods, convening advisory groups, and holding seminars and workshops. DOT is also using technology to increase public participation. In advance of finalizing its most recent environmental justice strategy, the department held traditional public notice and comment periods but added to this approach with a novel electronic participatory mechanism called EJ Ideascale. This electronic comment feature allowed the public to contribute their ideas *and* also view and respond to public comments made by others.[11]

According to the department's website, a considerable number of comments have been received on their updated environmental justice strategy. The department's response to public comments are also available online.[12] Yet again, the public comments highlight the challenges of advancing environmental justice in a large-scale, multifunctional bureaucracy. Some public comments strongly support the need for increased harmonization and integration of a unified environmental justice strategy within the department, while others advocate for more flexibility to ensure that decision making can accurately take into account unique, place-based features. Under these conditions, implementing agencywide environmental justice strategies remain challenging: federal agencies receive virtually no guidance on how to proceed when a diverse public provides conflicting input.

Conclusion

There is tremendous variation across federal agencies; for instance, they have divergent missions, different professional norms, and diverse agency cultures. EPA operates not only through its national offices, but also through its ten regional offices, DOE functions through various offices, laboratories, and field sites, while DOT works through hundreds of state transportation departments, MPOs, and local transit authorities, each of which enjoys substantial decision-making discretion. Such an overlay of arrangements leads to considerable variations in policy approaches across jurisdictions, especially since cities, states, and regions vary widely in their environmental problems. Thus, in many instances, environmental justice must contend with a fragmented and decentralized policy process.

Our assessment is based on examining a range of existing government documents and electronic sources. We compare information from these sources to existing research on civic engagement, public participation, and social equity to examine how well federal agencies meet the participation challenges embedded in EO 12898. Overall, the environmental justice movement and this executive order have increased opportunities for public participation in federal decision making. But despite the increased opportunity, participation from minority and low-income communities remains uneven. Perhaps this should not be surprising, given what we know about both the challenges inherent in generating diverse participation, particularly in communities that have been historically disenfranchised, and the challenges of creating systemic change in large bureaucracies.

The three federal agencies examined here have all experienced some success in using public participation to address environmental justice concerns, and they have also faced significant challenges. These challenges stem from multiple sources: (1) a complex decision-making environment at the federal level, complicated by devolution and privatization; (2) the capacity of large public organizations to incorporate public participation is variable; (3) the technical nature of environmental problems, which demands some level of competence on the part of the affected public, and the capacity of citizens to engage in scientific and technical decision making; and (4) the general decline in public engagement over time.

While there is significant promise in using public participation to address environmental justice concerns, this promise hinges on developing and maintaining effective public participation mechanisms. Prior research highlights that effective participation is characterized by clear goals, adequate human and financial capital investment, consistent institutional commitment, and an ability for participation to affect change in all stages of decision making (Beierle and Cayford 2002; Charnley and Engelbert 2005; Dietz at al 2008; Leach 2006; Reed 2008; Sirianni 2009). In the years since EO 12898 was signed into law, the EPA, DOE, and DOT have experienced challenges across the range of characteristics that describe effective participatory mechanisms. In the years ahead, if federal agencies want to increase opportunities for public participation, they should consistently invest more staff time and agency resources in a range of community outreach efforts. This can help build trust and establish a framework for effective participation if environmental justice problems arise. Federal agencies face a diverse and decentralized political system. Building agile organizational capacity to partner with state and local governments will remain essential in the years ahead. Similarly, federal agencies addressing environmental justice challenges would be well served by building the technical capacity of affected communities. This provides the affected communities more opportunity to partner with government agencies when tackling environmental justice problems. Across all of these efforts to engage the public, it is also imperative to design systematic interventions to allow for clear identification of what works and why when addressing environmental challenges.

Notes

1. Memorandum of Understanding on Environmental Justice and Executive Order 12898 (August 2011).

2. The accomplishment reports can be found at http://www.epa.gov/compliance/environmentaljustice/resources/reports/actionplans.html.

3. http://www.epa.gov/air/care/basic.htm.

4.http://www.epa.gov/environmentaljustice/resources/publications/grants/ej_smgrants_emerging_tools_2nd_edition.pdf contains a full description of EJ Small Grant projects.

5. This example is drawn from the following transcript: http://www.epa.gov/environmentaljustice/multimedia/transcripts/2012-09-20-community-outreach-call.pdf.

6. See the DOE website: http://energy.gov/mission.

7. http://energy.gov/sites/prod/files/EJStrategy_EO12898.pdf.

8. See the DOT website: http://www.dot.gov/about.

9. http://www.fhwa.dot.gov/environment/environmental_justice/ej_at_dot/order_56102a.

10.http://www.fhwa.dot.gov/environment/environmental_justice/ej_at_dot/dot_ej_strategy.

11. http://www.fhwa.dot.gov/environment/environmental_justice/ej_at_dot/2011_implementation_report.

12. http://www.fhwa.dot.gov/environment/environmental_justice/ej_at_dot/dot_ej_strategy/public_comment/index.cfm.

References

Agyeman, Julian, and Briony Angus. 2003. The Role of Civic Environmentalism in the Pursuit of Sustainable Communities. *Journal of Environmental Planning and Management* 46 (3): 345–363.

Baber, Walter F., and Robert V. Bartlett. 2013. Green Political Ideas and Environmental Policy. In *The Oxford Handbook of U.S. Environmental Policy*, ed. Sheldon Kamieniecki and Michael E. Kraft, 48–66. New York: Oxford University Press.

Bardach, Eugene. 1998. *Getting Agencies to Work Together: The Practice and Theory of Managerial Craftsmanship.* Washington, DC: Brookings Institution Press.

Beierle, Thomas C. 1998. *Public Participation in Environmental Decisions: An Evaluation Framework Using Social Goals.* Washington, DC: Resources for the Future.

Beierle, Thomas C., and Jerry Cayford. 2002. *Democracy in Practice: Public Participation in Environmental Decisions.* Washington, DC: Resources for the Future.

Brehm, John, and Scott Gates. 2002. *Working, Shirking, and Sabotage: Bureaucratic Response to a Democratic Public.* Ann Arbor, MI: University of Michigan Press.

Brion, Denis J. 1988. An Essay on LULU, NIMBY, and the Problem of Distributive Justice. *Boston College Environmental Affairs Law Review* 15: 437–504.

Brulle, Robert J., and David N. Pellow. 2006. Environmental Justice: Human Health and Environmental Inequalities. *Annual Review of Public Health* 27: 103–124.

Buchanan, Sariyah S. 2010. Why Marginalized Communities Should Use Community Benefit Agreements as a Tool for Environmental Justice: Urban Renewal and Brownfield Redevelopment in Philadelphia, Pennsylvania. *Temple Journal of Science, Technology, & Environmental Law* 29: 31–52.

Bulkeley, Harriet, and Arthur P. J. Mol. 2003. Participation and Environmental Governance: Consensus, Ambivalence, and Debate. *Environmental Values* 12 (2): 143–154.

Cairns, Shannon, Jessica Greig, and Martin Wachs. 2003. *Environmental Justice & Transportation: A Citizen's Handbook.* Berkeley, CA: ITS Berkeley.

Charnley, Susan, and Bruce Engelbert. 2005. Evaluating Public Participation in Environmental Decision-Making: EPA's Superfund Community Involvement Program. *Journal of Environmental Management* 77 (3): 165–182.

Coglianese, Cary. 1997. Assessing Consensus: The Promise and Performance of Negotiated Rulemaking. *Duke Law Journal* 46 (6): 1255–1349.

Coglianese, Cary. 2006. Citizen Participation in Rulemaking: Past, Present, and Future. *Duke Law Journal* 55 (5): 943–968.

Cole, Luke W., and Shelia R. Foster. 2001. *From the Ground Up: Environmental Racism and the Rise of the Environmental Justice Movement.* New York: New York University Press.

Commission on Civil Rights (CCR). 2003. *Not In My Backyard: Executive Order 12,898 and Title VI as Tools for Achieving Environmental Justice.* http://www.usccr.gov/pubs/envjust/ej0104.pdf.

Daley, Dorothy M. 2013. Public Participation, Citizen Engagement, and Environmental Decision Making. In *The Oxford Handbook of U.S. Environmental Policy,* eds. Sheldon Kamieniecki and Michael E. Kraft, 487–503. New York: Oxford University Press.

Department of Energy (DOE). 1995. *Environmental Justice Strategy: U.S. Department of Energy.* http://energy.gov/sites/prod/files/nepapub/nepa_documents/Red-Dont/G-DOE-EJ_Strategy.pdf.

Department of Energy (DOE). 2009. *Environmental Justice Five-Year Implementation Plan: First Annual Progress Report.* http://energy.gov/sites/prod/files/EJ_Progress_Report.pdf.

Department of Energy (DOE). 2011. *Environmental Justice Five-Year Implementation Plan: Second Annual Progress Report.* http://energy.gov/sites/prod/files/20110825%20EJ%20report%20WEB.pdf .

Department of Transportation (DOT). 1995. *Department of Transportation Environmental Justice Strategy.* http://www.epa.gov/environmentaljustice/resources/publications/interagency/dot-strategy-1995.pdf.

Department of Transportation (DOT). 1999. *Implementing Title VI Requirements in Metropolitan and Statewide Planning.* http://www.fhwa.dot.gov/environment/environmental_justice/facts/ej-10-7.cfm.

Department of Transportation (DOT). 2012. *Department of Transportation Environmental Justice Strategy.* http://www.fhwa.dot.gov/environment/environmental_justice/ej_at_dot/dot_ej_strategy.

Dietz, Thomas, Paul C. Stern, and the National Research Council. 2008. *Panel on Public Participation in Environmental Assessment and Decision Making, and National Research Council (U.S.) Committee on the Human Dimensions of Global Change. Public participation in Environmental Assessment and Decision Making.* Washington, DC: National Academies Press.

Environmental Protection Agency (EPA). 1995. *The EPA's Environmental Justice Strategy.* http://www.epa.gov/environmentaljustice/resources/policy/ej_strategy_1995.pdf.

Environmental Protection Agency (EPA). 1997. *Environmental Justice: 1996 Annual Report Working Toward Solutions.* Document no. EPA/300-R-97-004.

Environmental Protection Agency (EPA). 2004. *EPA Needs to Consistently Implement the Intent of the Executive Order on Environmental Justice.* Document no. 2004-P-00007. http://www.epa.gov/oig/reports/2004/20040301-2004-P-00007.pdf.

Department of Energy (DOE). 2008. *Final Site-Wide Environmental Impact Statement for Continued Operation for Los Alamos, New Mexico.* http://energy.gov/sites/prod/files/EIS-0380-FEIS-Summary-2008.pdf.

Environmental Protection Agency (EPA). 2011. *Plan EJ 2014.* http://www.epa.gov/environmentaljustice/resources/policy/plan-ej-2014/plan-ej-2011-09.pdf.

Environmental Protection Agency (EPA). 2013. *Plan EJ 2014 Progress Report.* Document no. EPA/300-R-13-001.

Fine, James D., and Dave Owen. 2005. Technocracy and Democracy: Conflicts between Models and Participation in Environmental Law and Planning. *Hastings Law Journal* 56 (5): 901–982.

Fletcher, Thomas H. 2003. *From Love Canal to Environmental Justice: The Politics of Hazardous Waste on the Canada-U.S. Border.* Peterborough, Canada: Broadview Press.

Foreman, Christopher H. 1998. *The Promise and Peril of Environmental Justice.* Washington, DC: Brookings Institution Press.

Furlong, Scott R. 2007. Business and the Environment: Influencing Agency Policymaking. In *Business and Environmental Policy: Corporate Interests in the American Political System*, ed. Michael E. Kraft and Sheldon Kamieniecki, 155–184. Cambridge, MA: MIT Press.

Gauna, Eileen. 1995. Federal Environmental Citizen Provisions: Obstacles and Incentives on the Road to Environmental Justice. *Ecology Law Quarterly* 22:1–87.

Government Accountability Office (GAO). 2011. *Environmental Justice: EPA Needs to Take Additional Actions to Help Ensure Effective Implementation.* Document no. GOA-12-77

Hansell, William H., Elizabeth Hollander, and Dewitt John. 2009. *Putting Community First: A Promising Approach to Federal Collaboration for Environmental Improvement.* Washington, DC: National Academy of Public Administration.

Hernandez, Willie G. 1995. Environmental justice: Looking Beyond Executive Order No. 12,898. *UCLA Journal of Environmental Law & Policy* 14:181.

Innes, Judith E., and David E. Booher. 2004. Reframing Public Participation: Strategies for the 21st Century. *Planning Theory & Practice* 5 (4): 419–436.

Kamieniecki, Sheldon. 2006. *Corporate America and Environmental Policy: How Often Does Business Get Its Way?* Stanford, CA: Stanford University Press.

Kellogg, Wendy A., and Anjali Mathur. 2003. Environmental Justice and Information Technologies: Overcoming the Information-Access Paradox in Urban Communities. *Public Administration Review* 63 (5): 573–585.

Klyza, Christoper McGrory, and David J. Sousa. 2013. *American Environmental Policy: Beyond Gridlock.* Cambridge, MA: MIT Press.

Kraft, Michael E., and Bruce B. Clary. 1991. Citizen Participation and the NIMBY Syndrome: Public Response to Radioactive Waste Disposal. *Western Political Quarterly* 44 (2): 299–328.

Larson, Kelli L., and Denise Lach. 2008. Participants and Non-Participants of Place-Based Groups: An Assessment of Attitudes and Implications for Public Participation in Water Resource Management. *Journal of Environmental Management* 88 (4): 817–830.

Laurian, Lucie. 2007. Deliberative Planning through Citizen Advisory Boards— Five Case Studies from Military and Civilian Environmental Cleanups. *Journal of Planning Education and Research* 26 (4): 415–434.

Leach, William D. 2006. Collaborative Public Management and Democracy: Evidence from Western Watershed Partnerships. *Public Administration Review* 66 (1): 100–110.

Lubell, Mark, Paul A. Sabatier, Arnold Vedlitz, Will Focht, Zev Trachtenberg, and Marty Matlock. 2005. Conclusions and Recommendations. In *Swimming Upstream: Collaborative Approaches to Watershed Management*, ed. Paul A. Sabatier, Will Focht, Mark Lubell, Zev Trachtenberg, Arnold Vedlitz, and Marty Matlock, 261–296. Cambridge, MA: MIT Press.

Macedo, Stephen. 2005. *Democracy at Risk: How Political Choices Undermine Citizen Participation and What We Can Do about It.* Washington, DC: Brookings Institution Press.

Mazmanian, Daniel A., and Paul A. Sabatier. 1989. *Implementation and Public Policy: With a New Postscript.* Lanham, MD: University Press of America.

Mohai, Paul, David Pellow, and J. Timmons Roberts. 2009. Environmental Justice. *Annual Review of Environment and Resources* 34: 405–430.

Munton, Don, ed. 1996. *Hazardous Waste Siting and Democratic Choice.* Washington, DC: Georgetown University Press.

National Environmental Justice Advisory Council (NEJAC). 2004. *Environmental Justice and Federal Facilities: Recommendations for Improving Stakeholder Relations between Federal Facilities and Environmental Justice Communities.* http://www.epa.gov/compliance/ej/resources/publications/nejac/ffwg-final-rpt-102504.pdf.

Pellow, David N., and Robert J. Brulle. 2005. *Power, Justice, and the Environment: A Critical Appraisal of the Environmental Justice Movement.* Cambridge, MA: MIT Press.

Pressman, Jeffrey L., and Aaron B. Wildavsky. 1984. *Implementation: How Great Expectations in Washington Are Dashed in Oakland: or, Why It's Amazing that Federal Programs Work At All, This Being A Saga of the Economic Development Administration as Told by Two Sympathetic Observers Who Seek to Build Morals on a Foundation of Ruined Hopes.* 3rd ed. Berkeley, CA: University of California Press.

Putnam, Robert D. 2000. *Bowling Alone: The Collapse and Revival of American Community.* New York: Simon & Schuster.

Reed, Mark S. 2008. Stakeholder Participation for Environmental Management: A Literature Review. *Biological Conservation* 141 (10): 2417–2431.

Rosenbaum, Walter A. 2002. *Environmental Politics and Policy.* 5th ed. Washington, DC: CQ Press.

Rosenstone, Steve J., and John M. Hansen. 1993. *Mobilization, Participation, and Democracy in America.* New York: Longman.

Sander, Thomas H., and Robert D. Putnam. 2010. Still Bowling Alone? The Post-9/11 Split. *Journal of Democracy* 21 (1): 9–16.

Schlozman, Kay L., Sidney Verba, and Henry E. Brady. 2012. *The Unheavenly Chorus: Unequal Political Voice and the Broken Promise of American Democracy.* Princeton, NJ: Princeton University Press.

Sirianni, Carmen. 2009. *Investing in Democracy: Engaging Citizens in Collaborative Governance.* Washington, DC: Brookings Institution Press.

Spyke, Nancy Perkins. 1999. Public Participation in Environmental Decisionmaking at the New Millennium: Structuring New Spheres of Public Influence. *Boston College Environmental Affairs Law Review* 26: 263–314.

Vajjhala, Shalini P. 2010. Building Community Capacity? Mapping the Scope and Impacts of EPA's Environmental Justice Small Grants Program. *Research in Social Problems and Public Policy* 18: 353–381.

Weber, Edward P. 2003. *Bringing Society Back In: Grassroots Ecosystem Management, Accountability, and Sustainable Communities*. Cambridge, Mass.: MIT Press.

Webler, Thomas, and Seth Tuler. 2006. Four Perspectives on Public Participation Process in Environmental Assessment and Decision Making: Combined Results from 10 Case Studies. *Policy Studies Journal: the Journal of the Policy Studies Organization* 34 (4): 699–722.

Wilson, James Q. 1989. *Bureaucracy: What Government Agencies Do and Why They Do It*. New York: Basic Books.

Yackee, Jason Webb, and Susan Webb Yackee. 2006. A Bias towards Business? Assessing Interest Group Influence on the U.S. Bureaucracy. *Journal of Politics* 68 (1): 128–139.

7

Evaluating Fairness in Environmental Regulatory Enforcement

David M. Konisky and Christopher Reenock

In November 2011, investigative journalists from the Center for Public Integrity (CPI) and National Public Radio (NPR) completed a series of reports, entitled *Poisoned Places*, which chronicled the experience of several communities confronting exposure to hazardous substances from large industrial air pollution sources (CPI/NPR 2011). In their reporting, the journalists uncovered an internal "watch list" compiled by the Environmental Protection Agency (EPA), which included hundreds of facilities, most of which were high-priority violators (HPVs) of the Clean Air Act (CAA).[1] HPV facilities are major air polluters, such as large manufacturing facilities and power plants, that are failing to meet core CAA obligations, usually pollution performance standards. From the EPA's perspective, the watch list is a management tool to help the agency identify: "recidivist and chronically noncomplying facilities whose violations have not been formally addressed by either the state or the EPA" (EPA 2008). From an outsider's perspective, the list reads as an inventory of the "worst-of the-worst" sources of illegal air pollution, or, alternatively, a list of facilities that have not received sufficient attention from the government agencies responsible for enforcing the CAA. Of course, both could be true as well.

The watch list published in July 2012 included 340 facilities, more than half of which were in just four states (33 in Illinois, 62 in Louisiana, 48 in Ohio, and 32 in Texas). A quick analysis suggests that these facilities were often located in poor and minority communities. Whereas the average percentage of African-Americans and Hispanics in ZIP codes across the country is 7.7 percent and 8.7 percent, respectively, the averages based on the location of the watch-list facilities were 15.7 percent and 11.8 percent, respectively. The percentage of the communities living in poverty was about 2.5 points above the national average for facilities included on the watch list.[2]

The CPI/NPR team concluded that a major part of the problem was that the EPA and state environmental agencies were failing to pursue aggressive enforcement of these repeat offenders, and as a result, the communities living nearby were suffering from severe pollution burdens with potentially significant adverse health effects. To the extent to which these communities were low-income, minority, or both, this might reflect an environmental injustice. In fact, in one of the segments of the report, a high-ranking EPA official acknowledged as much. The segment featured a local community's efforts to bring attention to the toxic emissions from Tonawanda Coke, a foundry coke and coal tar manufacturer located in Tonawanda, New York. For years, community residents complained to federal and state officials about what they believed were excessive and illegal emissions of benzene and other chemicals from the plant. In addressing the slow government response, Judith Enck, administrator of the EPA regional office with jurisdiction over New York, said, "If this was in an affluent city where thousands of people lived, I think there would have been more of a laser-like focus on this earlier" (Shogren, Lombard, and Bartlett 2011). This is a revealing statement from such a high-ranking EPA official. Given the fact that Tonawanda is not a prototypical "environmental justice" community (it is 98 percent white and economically working class), it raises the question of whether this example of government inattention is more egregious in minority and poor areas.

To be sure, environmental justice advocates have long alleged that minority and low-income communities experience disproportionate environmental hazards, in part, because of biased enforcement of pollution control and public health laws (Bryant 1995; Bullard 1993; Bullard & Johnson 2000; Collin 1993). A couple of statements from Robert Bullard are illustrative: "Institutional racism influences decisions on local land use, *enforcement of environmental regulations*, industrial facility siting, management of economic vulnerability . . ." (emphasis added) (Bullard 1993). And, in a statement quite similar to that made by Enck (although it came twenty years earlier), Bullard commented: "People say decisions are made based on risk assessment and science. The science may be present, but when it comes to implementation and policy, a lot of decisions appear to be based on the politics of what's appropriate for that community. And low-income and minority communities are not given the same priority, nor do they see the same speed at which something is perceived as a danger and a threat" (quoted in Lavelle and Coyle 1992).

Do EPA and state enforcement officials less vigorously enforce pollution control laws when facilities are located in poor and minority communities? Specifically, do they conduct less compliance monitoring, impose lighter administrative sanctions, or both with firms that have been found to be violating major environmental statutes? And did the environmental justice policies adopted in the early 1990s result in fewer disparities of this sort?

We address these and other related questions in this chapter. We begin with a review of existing studies that examine inequities in environmental enforcement, and then describe the federal government's policy response in the area of compliance assurance and enforcement. We then conduct original statistical analysis of enforcement of the federal CAA to determine if this policy response resulted in beneficial changes in government efforts at both the federal and state level. Overall, we find that the policy reforms did not create large changes to EPA or state enforcement efforts, and in some cases, disparities seemed to worsen. We conclude that, at least in the case of the CAA, the policy attention to equity concerns did not generate a shift in enforcement attention to facilities located in poor and minority communities. In the concluding section of the chapter, we discuss the implications of these findings and evaluate whether recent EPA policy efforts are likely to be more effective in the future.

Past Research on Enforcement Disparities

Over the past twenty-five years, social scientists have devoted significant effort to identifying and quantifying the degree to which poor and minority communities experience disproportionate environmental burdens. This literature, summarized in chapter 1 and referenced throughout the book, has become increasingly sophisticated, and there has accumulated considerable evidence of race- and class-based environmental burdens in facility location and exposure to pollution (Ringquist 2005).

Much less attention has been given to the question of whether there are systematic race- and class-based disparities in the enforcement of environmental laws. This is somewhat surprising given the repeated claims that poor and minority communities are often overlooked by regulators. As is discussed in more detail in chapter 8, some studies have examined outcomes from judicial proceedings in cases where polluters were penalized for violating major environmental statutes. A study by Lavelle and Coyle (1992), published in the *National Law Journal*, concluded that there were significant disparities. Examining federal district court decisions from 1985 to 1991 that involved violations of air pollution, water pollution, and hazardous waste laws, Lavelle and Coyle concluded that, on average, fines were about $50,000 lower when facilities were located in minority and poor areas. These findings were, and continue to be, widely cited by environmental justice advocates.

Subsequent research, however, has concluded that the Lavelle and Coyle study was severely flawed. As pointed out by Evan Ringquist (1998), the study selected a short time frame (with no clear rationale), computed average penalties across vastly different types of programs, failed to include control variables to account for factors other than demographics that might explain outcomes, and finally, did not employ tests of statistical significance. In an original analysis that addressed these concerns, Ringquist (1998) found no real race- or class-based disparities in either average penalty amounts in total or in average penalty amounts per violation. In fact, over some time periods that he considered, the average penalty amounts were *higher* for facilities in low-income and minority areas.

Studies of judicial outcomes examine decisions made at the end of what can be a very long regulatory enforcement process. It can take many years for a case where a firm has been accused of committing major violations of a statute such as the CAA or the Clean Water Act (CWA) to reach a judicial decision. In fact, most cases never reach this stage; instead, they are resolved through out-of-court negotiated settlements, which are then usually formalized in consent decrees. Thus, even if there are no race- or class-based disparities in judicial decisions, such disparities may still exist at earlier stages of the enforcement process. For example, government inspectors may pursue less intense compliance monitoring of regulated facilities when they are located in poor and minority communities. Inspections are the principal means by which government agencies detect noncompliant behavior, and the EPA has long pursued a deterrence strategy to discourage regulated entities from violating their environmental obligations (Rechtschaffen and Markell 2003). Similarly, government agencies may elect to impose less severe administrative sanctions against noncompliant firms in these communities. That is, even if agencies detect significant violations, they may treat similar violations differently. Such unequal enforcement may, at least in part, explain observed inequitable patterns of facility location and pollution burdens. For example, firms may be more inclined to site a new facility in a given jurisdiction if a government agency has a reputation for lenient enforcement in that area. Moreover, less stringent enforcement of facilities in poor and minority areas may contribute to higher observed levels of pollution in these communities if facilities in these areas are failing to comply with their pollution control obligations.

We have found some evidence in our previous work of race- and class-based inequities in regulatory enforcement. Studying state enforcement of the CAA, CWA, and the Resource Conservation and Recovery Act (RCRA), Konisky (2009a) found that over the period between 1985 and 2000, states performed fewer inspections and imposed fewer punitive actions in poor and low-income counties. Similar differences did not emerge regarding race or ethnicity. Konisky and Schario (2010) investigated whether the pattern of disparities identified at the county level were also present when bringing the unit of analysis down to the facility level. Studying large water pollution sources, they found mixed evidence. In some of their regression models, they estimated lower likelihoods of inspections for facilities in large poor and Hispanic communities, but *higher* likelihoods of inspections directed at facilities in large African-American communities. Results were similarly mixed for administrative sanctions. In Konisky and Reenock (2013b), we found that facilities located in Hispanic and lower-class communities were both more likely to be significant violators of the CAA and less likely to be characterized as such by state regulatory officers.

Numerous other studies of regulatory enforcement have tested for race- and class-based disparities, with varied results. Several studies have found that inspections of facilities regulated under the CWA are less likely when the facilities are located in an area with a high percentage of low-income populations (Earnhart 2004a, b; Helland 1998; Scholz and Wang 2006), but others have found that facilities in poorer areas are characterized by more regulatory activity (Gray and Shadbegian 2004, 2012). With respect to punitive actions, some studies have found that facilities in areas with large poor populations tend to be associated with fewer punitive enforcement measures under both the CAA and the CWA (Gray and Shadbegian 2004), but other research does not reveal statistically significant differences (Gray and Shadbegian 2012). With respect to race and ethnicity, one study estimated a negative association between the percentage of African-American and Hispanic residents in an area and the likelihood of a CWA inspection (Scholz and Wang 2006). Other studies have found that CWA-regulated facilities located in areas with more minorities are not differentially targeted by enforcement officials (Gray and Shadbegian 2012), and may in fact see more enforcement (Gray and Shadbegian 2012).

In sum, the evidence is mixed. The varied results may be explained by any number of reasons, ranging from the heterogeneity in programs and time periods studied to data measurement and statistical modeling strategies. Moreover, even if the studies finding disparities are correct, one cannot conclude with certainty that they are due to overt government discrimination. Although one cannot rule out this possibility, it is not the only plausible explanation. Several scholars have suggested that the likely mechanism at work is political capacity (Gray and Shadbegian 2012; Hamilton 1995; Hamilton and Viscusi 1999; Konisky and Reenock 2013b). Firms located in communities that can effectively overcome collective action problems and advocate for strong enforcement are more likely to demand and secure government attention toward regulated entities in their areas. We know from decades of research in political science that minorities and individuals with low socioeconomic status tend to participate less in the political process through actions such as voting, signing petitions, attending local meetings, or writing letters to members of Congress (e.g., Leighley and Vedlitz 1999; Rosenstone and Hansen 1993; Verba and Nie 1972; Verba, Schlozman, and Brady 1995). Communities consisting of large numbers of poor and minority citizens, thus, may be less likely to mobilize and apply pressure on government agencies. It is also important to note, however, that when these communities do successfully mobilize, they can mitigate some of these effects. In our recent study of the CAA from 2001 to 2004 (Konisky and Reenock 2013b), we found that facilities are less likely to violate the law and regulatory officials are more likely to detect compliance in communities with more local environmental justice advocacy organizations.

Sorting out the mechanisms behind patterns of enforcement inequities is beyond the scope of this chapter. Our focus instead is on the question of whether government enforcement efforts changed as a response of the policies put in place in the early 1990s to address concerns that pollution control laws were not being equally enforced. Before we move to this analysis, we first describe these policies and identify their possible influence on EPA and state enforcement priorities.

The Federal Policy Response

In response to the growing evidence of environmental inequities, an increasingly active and impatient environmental justice movement began to demand a federal policy response. As has been described in earlier chapters of this book, administrative reforms, such as Executive Order 12898 on Environmental Justice (EO 12898) signed by President Bill Clinton in 1994, sought to bring attention to environmental justice throughout the decision making of the federal bureaucracy, particularly the EPA. The reforms specifically endorsed the use of enforcement tools as a remedy for environmental inequities. EO 12898, for instance, directed federal agencies to "promote enforcement of all health and environmental statutes in areas with minority populations and low-income populations." The Environmental Justice Strategy developed by the EPA in 1995 declared the need to work with state governments to "identify and respond to any regulatory gaps in the protection of covered populations" and, more generally, address environmental equity concerns in their regulatory and enforcement programs (Browner 1995). In addition, the National Environmental Justice Advisory Council (NEJAC) called on the EPA and the states to strengthen their enforcement efforts in minority and low-income communities (NEJAC 1995). And, finally, EPA lawyers in 1994 determined that the agency had "broad authority" under existing environmental statutes to address environmental justice issues through enforcement measures (Kuehn 2000).

Despite the explicit endorsement of an environmental justice–based enforcement strategy, there has been almost no scholarship studying its efficacy (Konisky 2009b is an exception). Did these administrative actions lead to more intense surveillance and stricter penalties for facilities located in poor and minority communities? The answer to this question is not obvious. On one hand, EO 12898 and the other administrative reforms put in place in the mid-1990s did not carry the force of law. The EPA and state agencies responsible for enforcement of federal environmental laws were (and remain today) under no legal obligation to shift compliance monitoring resources to facilities located in poor and minority communities, or to ratchet up the severity of punitive sanctions at these facilities when they uncover violations. The policy initiatives put in place during the early 1990s, therefore, represent a relatively weak policy intervention, and there may not be a strong reason to expect that they resulted in a major shift in regulatory enforcement efforts. On the other hand, the EPA was under considerable pressure from the environmental justice community to take equity concerns seriously. It is certainly possible that the new initiatives sent a strong signal from the highest levels of the Clinton administration that they expected the EPA to act, including through enforcement actions.

Moreover, the potential effects do not end with the EPA. Much of the U.S pollution control system is designed such that states are either required (e.g., CAA State Implementation Plans) or can request permission (e.g., CWA NPDES permit programs) to implement and enforce major programs. EPA delegates program *primacy* to states that have demonstrated that they have at least as stringent regulatory programs enacted into state law. The federal government carries out oversight of state efforts and retains the right to withdraw primacy if it believes that state efforts are failing to meet federal standards (the EPA operates programs in states without primacy). Within this system, states have extraordinary discretion to choose how much effort they would like to exert toward enforcing federal programs (Konisky 2007; Sigman 2003).

The relevant question here is whether federal environmental justice policy led to a change in state enforcement activity. This may have happened in an indirect way. By raising the profile of equity concerns, the Clinton administration made environmental justice a clear priority. Even if these initiatives were only symbolic, they may still have led states to make equity an important part of their own agendas. The EPA also attempted to more directly influence the states through the use of financial mechanisms. The agency, for example, created a grants program to provide financial assistance to state and tribal governments interested in taking actions to reduce inequities or to build institutional capacity to address environmental justice issues. The EPA also made federal funding to states contingent on their compliance with the agency's regulations regarding Title VI of the Civil Rights Act, which extended to state permitting of facilities regulated under federal pollution control laws for those states delegated the authority to implement these programs (Ringquist and Clark 1999).

Evaluating Policy Change

Before we proceed to testing for the presence of policy effects in the context of enforcement of the CAA, it is important to specify the standards that we will use to evaluate them. Assessing policy effects in this case is less straightforward than it might at first seem. The reason is that we do not know ex ante what enforcement patterns existed during the pre-policy period. It is often assumed that the pre-policy period was characterized by extreme disparities in EPA and state compliance monitoring, punitive sanctions, or both. This is reflected in assertions from environmental justice advocates and from some researchers writing in the academic literature, and was at least implied by the language of EO 12898 itself. But it is also possible that no such disparities existed in the pre-policy period. It is even possible that facilities in poor and minority communities were already subject to more intense enforcement activity before the environmental justice reforms were put in place. As noted previously, there is no consensus in previous empirical studies to rely on. Because of the multiple possible baseline conditions, there is no single test from which we can conclude that the reforms had a beneficial effect, no effect, or perhaps even a detrimental effect.

In the analysis that follows, we examine the presence of disparities in both the pre- and post-policy periods more systematically, taking into consideration the multitude of factors related to facility-level enforcement decisions. We will refer to two types of disparities: negative and positive. Negative disparities are cases where government agencies performed relatively fewer enforcement efforts in environmental justice target communities than in nontarget communities (we define these communities next). Positive disparities are cases where government agencies conducted relatively more enforcement in these target communities. The other possible outcome, of course, is no disparities, in which case government agencies are carrying out enforcement similarly across these communities. The case of no disparities suggests that facilities in low-income and minority communities are not being systematically neglected, at least compared to wealthier and nonminority communities, when it comes to regulatory enforcement. In the next sections, we summarize the results of these tests (twelve in total) for each combination of agency (EPA or state), action (inspection or punitive action), and environmental justice community (African American, Hispanic, or poor).

Analyzing Policy Effects in the Clean Air Act

The environmental justice initiatives coming out of the Clinton administration during the first half of the 1990s were not program-specific. Rather, they aimed to raise the profile of environmental inequities throughout the activities of the EPA, as well as other federal agencies. Given our focus on enforcement of pollution control laws, it is appropriate to consider a major federal statute such as the CAA.

Research Design

To analyze the effects of federal environmental policy on enforcement, we used a simple interrupted time series research design. Interrupted time series is a quasi-experimental method that is frequently used to analyze the effects of a policy intervention (see Shadish, Cook, and Campbell 2002 for an overview). Specifically, we estimated a statistical model that examines EPA and state regulatory enforcement of the CAA as a function of (1) a set of demographic characteristics that identify environmental justice communities; (2) a policy variable that signifies the period before and after the environmental justice policies were put in place; and (3) a group of control variables to capture other correlates of regulatory enforcement outcomes.[3] Of central interest in this model were the tests of the effects of the policy intervention on both EPA and state enforcement behavior in facilities located in environmental justice communities.[4]

Data

To investigate whether federal and state regulatory enforcement patterns changed in response to the federal policies put in place during the mid-1990s, we needed to first identify a set of regulated facilities. In the

analysis described next, we considered all currently active major air polluters regulated by the CAA.[5] Major sources of air pollution are required
to obtain a Title V operating permit; they typically include facilities that
have the potential to emit at least 100 tons per year of any criteria air
pollutant. Major air sources also include facilities that emit hazardous air
pollutants above certain thresholds as determined by EPA guidelines. In
the analysis that follows, there were about 12,800 such facilities in the
United States.[6]

Enforcement Measures   The dependent variables we analyzed are CAA
enforcement actions taken by either the EPA or state governments. Historical enforcement data are available from the EPA's Integrated Data for
Enforcement Analysis (IDEA) database, from which we created four separate measures: EPA inspections, EPA punitive actions, state inspections,
and state punitive actions. For each major air source, we created a dichotomous measure of whether the EPA (or state) performed an inspection or
took a punitive action against it in a given year (years in which an action
was recorded were coded as 1, and years without were coded as zero).
Inspections are the principal actions taken by the EPA and state governments to determine the compliance status of regulated facilities, and they
usually (although not always) include emissions tests and assessments of
pollution control technologies. *Punitive actions* include measures taken
to bring noncompliant firms back into compliance, which include both
informal actions (such as notifications of violation) and formal actions
(such as administrative orders, consent decrees, and civil penalties). The
unit of analysis, therefore, was a facility-year, and we considered the
period from 1990 to 2008. (Summary statistics for all variables are displayed in table A7.1 in the appendix.)

Because our measures of enforcement activities were binary (either an
action was taken at facility in a given year or it was not), we used a logistic regression model, and we did so separately for EPA inspections, EPA
punitive actions, state inspections, and state punitive actions.

Environmental Justice Communities   We employed a standard suite of
demographic measures to characterize communities. Of most importance
are three variables we used to distinguish communities that we would
expect to have been targeted by the EPA and state governments for

additional enforcement attention in light of the federal environmental justice initiatives. To define these "environmental justice communities," we first located each of the 12,800 major air sources in geographical space using latitude and longitude information from the EPA's Geospatial Data Access Project. Then, to measure the demographic characteristics of the population living around each of these facilities, we used an areal apportionment method (Mohai and Saha 2006, 2007; Konisky and Schario 2010).[7] Specifically, we used geographic information system (GIS) software to identify the demographic information for people living within 1 mile of each facility.[8] We did this separately for 1990, 2000, and 2010, and then we employed linear interpolation to impute the values for the inter-decennial census years. The output, then, was an annual measure of the percentage of the population living around each facility that was poor, African-American, and Hispanic.[9]

There remains no agreed-upon standard for what constitutes a "poor" or "minority" community in the environmental justice context. In the analysis that follows, we defined target communities as ones in the top decile of African-American, Hispanic, and poverty populations, based on the measurement of demographics discussed previously.[10] Using this threshold, the target communities were those with 39 percent, 32 percent, and 27 percent or more of their populations being African-American, Hispanic, or poor, respectively.[11] Within these communities, the population demographic statistics for each targeted group had mean values of 58 percent, 52 percent, and 34 percent, with standard deviations of 15 percent, 16 percent, and 6.6 percent, respectively. Designating target communities in this way reflects the approach stated by the EPA in its environmental justice strategy that it would direct more enforcement attention to poor and minority communities.

Policy Intervention To capture the federal policy intervention, we created a dummy variable coded 1 for the years 1995 and later, and 0 for prior years. Using 1995 as our policy cutoff point allowed us to test for the effects of the federal environmental justice initiatives during the period after the policies were announced.[12] Because we were interested in whether the policy led to enhanced enforcement effort in target communities, we created a two-way interaction term for each community indicator and the policy intervention variable.

Control Variables We included a variety of control variables in our regression analysis to guard against making incorrect inferences. First, we included several additional demographic variables, including continuous measures of the percentage of African Americans, Hispanics, and the poor in the population; median household income; and the percentage of the tract population with at least a high school education. These were measured using the same procedure as described previously for the community demographic attributes. In addition, we controlled for possible heterogeneity across firms. We included controls for different types of major air polluters by creating a series of dummy variables representing different industrial sectors: utilities, manufacturing, mining, and oil and gas.

Our models also included measures of air pollution severity in the area in which the facility is located, in addition to economic and political conditions. To capture pollution severity, we used information from the EPA *Green Book* on county-level nonattainment status. Each year, the EPA designates every county in the United States as in either in attainment (full or partial) or nonattainment with National Ambient Air Quality Standards (NAAQS) for six criteria air pollutants. The measure we used was a simple dichotomous variable of whether the county in which the major air source was located was in nonattainment for at least one of the six pollutants. In terms of economic conditions, we used the county-level unemployment rate derived from Bureau of Labor Statistics data. State political conditions are measured with three variables: an indicator variable for whether the governor of the state was a Democrat (coded 1 for yes, and zero for no), the percentage of Democrats in the state legislature across both chambers, and state citizen ideology using the revised 1960–2008 citizen ideology series developed by Berry et al. (1998). Finally, the regression models also included dummy variables for presidential administrations to account for national-level differences that come with different EPA regimes, dummy variables for each of the ten EPA regional offices, and lagged (by one year) measures of facility-level enforcement actions.

Did CAA Enforcement Change after the Policy Reforms?

We focus the discussion of our statistical findings on the question of whether the EPA and states directed more enforcement attention to poor and minority communities in the post-policy period. First, we briefly note that the models themselves perform reasonably well in predicting the relevant enforcement action.[13] In addition, on the whole, the control variables (whether task factors, political features, or firm-level characteristics) relate to the relevant actions in line with expectations from the literature.[14] State actions, particularly punitive actions, appear to be, on average, more sensitive to local- and state- level factors compared to EPA actions.

The complete regression results are reported in table 7A.2 in the appendix. Here, we present the model estimates of the differences in the predicted probability of inspections and punitive actions for a given target community relative to a nontarget community. Predicted probabilities provide a useful way to assess the substantive effects from our statistical model. We do this for both the pre-policy period and the post-policy period, since we are interested in determining changes in the overall pattern of enforcement directed at facilities in African-American, Hispanic, and poor communities relative to the rest of the communities where facilities are located. Figure 7.1 displays these quantities for EPA enforcement actions, and figure 7.2 displays them for state enforcement actions. In the graphs, positive disparities are cases where the 95 percent confidence interval is fully above zero, negative disparities are cases where the 95 percent confidence interval is fully below zero, and no disparities are cases where the 95 percent confidence interval includes zero.

Beginning with the EPA, our analysis indicates that, in the post-policy period, the EPA was more likely to inspect firms in African-American communities but less likely to inspect facilities in Hispanic communities compared to all other communities. There was no difference for poor communities. These results are displayed in the top panel of figure 7.1. With respect to EPA performance on punitive actions (bottom panel of figure 7.1), all of the tests in the post-reform era suggest no differential treatment across environmental justice target and nontarget communities.

To fully assess any policy effects, it is also necessary to examine patterns of enforcement during the pre-policy period. Doing so is revealing in a couple of ways. First, we find that there was a higher likelihood of an EPA inspection at facilities in an African-American community *even*

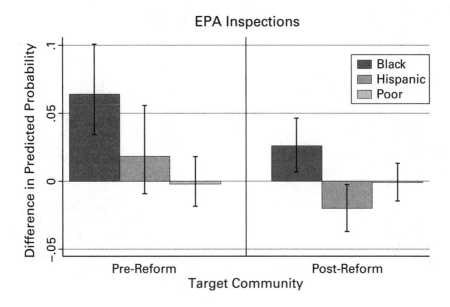

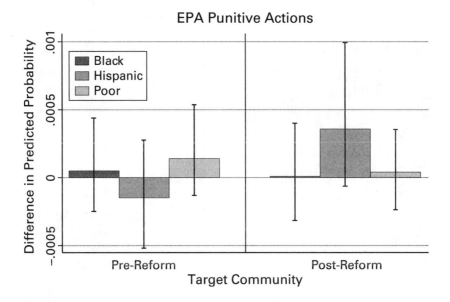

Figure 7.1

Impact of Policy Reforms on EPA CAA Enforcement. Note: Values represent the estimated predicted probability of an action directed at a facility in a environmental justice community minus the estimated predicted probability of an action directed at a facility in all other communities. The bars represent the 95 percent confidence interval.

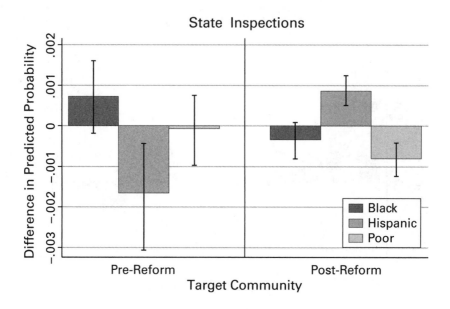

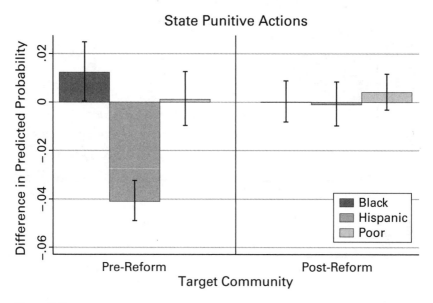

Figure 7.2

Impact of Policy Reforms on State CAA Enforcement. Note: Values represent the estimated predicted probability of an action directed at a facility in a environmental justice community minus the estimated predicted probability of an action directed at a facility in all other communities. The bars represent the 95 percent confidence interval.

before the policy reforms were put in place. In addition, in terms of punitive sanctions, the EPA was no less likely to impose a sanction at a violating firm in African-American, Hispanic, or poor communities before the reforms than it was in all other communities. In sum, for inspections and punitive actions, there were actually *no* preexisting negative disparities in EPA actions. The only difference in enforcement patterns between the pre- and post-policy period, thus, was for facilities in Hispanic communities, and in this case, these facilities actually fared *worse* in 1995 and beyond.

Switching attention to state agencies, we see a somewhat different pattern. With respect to inspection activity, as displayed in the top panel of figure 7.2, states were more likely to inspect a major air polluter in the post-policy period in Hispanic communities but less likely to do so at a facility located in a community with a high level of poverty or African-American residents. In the case of state punitive actions, the lower panel in figure 7.2 suggests no difference in the predicted probability of an action in any of the environmental justice target communities. At first glance, then, there does not seem to be much evidence that state agencies targeted more enforcement to poor and minority communities (with inspections in Hispanic areas being the lone exception) after receiving the policy signal from the federal government.

Again, however, we must also consider the patterns of state activity before the policy reforms. Doing so shows positive gains, but only with respect to Hispanic communities. From 1990 to 1994, states were less likely to perform inspections and carry out punitive sanctions at facilities located in Hispanic communities, whereas in the post-policy period, they were *more* likely to conduct inspections and *no less* likely to use sanctions. The results were more mixed for African-American and poor communities. In a couple of cases—inspections at facilities in African-American communities and punitive actions in poor communities—there is no evidence of different patterns before and after the environmental justice policies were put in place. However, in two other cases—inspections in poor communities and punitive actions in African-American communities—the evidence suggests that enforcement intensity may have actually *decreased* in the post-policy period. The magnitude of this shift for poor communities, while statistically significant, was negligible, shifting to a post-policy difference of a 0.0008 lower probability of inspecting firms in poor versus nonpoor communities. The change for African-American communities was larger, but nevertheless substantively small. Relative to firms in non-African-American communities, firms in African-American communities moved from a 0.012 pre-policy difference in the probability of receiving a punitive action to no difference.

To put all of these findings in context, the largest substantive effect associated with the EPA's environmental justice policies estimated here was for firms in Hispanic communities regarding state punitive actions. Our results suggest that prior to the policy, firms in non-Hispanic communities had an estimated probability of receiving a punitive action from a state agency of approximately 0.07, while firms in Hispanic communities faced an estimated probability of only 0.03 for receiving a similar action—a more than 57 percent difference. In the post-policy era, this difference was erased, shifting to the statistical equivalent of "no detectable difference." This fairly substantial policy effect greatly exceeds the more common effect sizes estimated for all other statistically significant findings. In short, outside of the policy effect for the Hispanic communities noted previously, the estimated substantive effects of the federal policy initiatives, when we detected any, were, on the whole, unremarkable.

To this point, we have assumed that the EPA and states identify environmental justice target communities solely on the basis of demographics. Given that one of the goals of the federal environmental justice policy reforms was to direct more government attention to poor and minority communities that were overburdened by pollution, we also conducted an analysis taking this into account. Specifically, we estimated the same set of regression models, but this time we used a three-way interaction term, where the environmental justice community and policy variable was also multiplied by total toxic emissions (aggregated to the ZIP code) using data from the EPA's Toxics Release Inventory (TRI). The coefficient on this term allows for our test of policy effects to be conditional on the level of toxic emissions in the area. We do not report the results here due to space limitations, but in brief, we did not find evidence that the EPA and states targeted facilities in high–toxic emissions areas any differently.

Putting the results of the analysis together, what can we say about the effects of the policy reforms on enforcement of the CAA? Recall from the discussion earlier in the chapter that we characterize a shift toward more enforcement as a beneficial outcome, a shift toward less enforcement as a detrimental outcome, and a case of little meaningful change as no difference. As summarized in table 7.1, of the twelve tests conducted in our main analysis, seven suggest that the policy had no impact, two suggest a beneficial impact, and three suggest a detrimental impact.

Table 7.1
Estimated Policy Outcomes

	Post-Policy Period		
Pre-Policy Period	*Negative Disparity*	*No Disparity*	*Positive Disparity*
Negative Disparity	0	1	1
No Disparity	2	6	0
Positive Disparity	0	1	1

Note: "Disparity" refers to estimated differences in a given environmental justice (EJ) community for a given expected agency action relative to non-EJ communities. A Positive Disparity exists when a given agency action is more likely for an EJ community relative to non-EJ communities. A Negative Disparity exists when a given agency action is less likely for an EJ community relative to non-EJ communities. No Disparity exists when a given agency action is equally likely in EJ and non-EJ communities.

It appears, therefore, that the policies put in place to address environmental justice had minimal impact—that is, by and large, there was not much discernible change in agency enforcement efforts in poor and minority communities. However, this does not necessarily mean that EPA enforcement was biased in these areas, as often alleged by environmental justice advocates. Rather, the EPA on the whole was already carrying out enforcement in target communities in a similar way as in nontarget communities. In fact, in the case of African-American communities, the agency was devoting more attention to major air polluters in these areas. The one exception to this general pattern is that enforcement effort did decline with respect to facilities in Hispanic communities, although we do not attribute this outcome to the policy reforms themselves. Moreover, the policy reforms do not appear to have had much of an effect on how state administrative agencies carried out their enforcement efforts. Although the reforms appear to have driven up enforcement of facilities in Hispanic communities, enforcement of facilities in African-American and poor communities at best stayed the same, but in some cases, it seemed to have become less intense.

Implications for Future Policy

An important question in any policy analysis or program evaluation is: What constitutes success? In some situations, the answer may be straightforward, but that is not necessarily the case in the context of environmental justice. For some people, enhanced levels of government attention to pollution sources located in poor and minority communities may qualify. According to this view, what is required is that these communities receive not just equal attention, but that they receive extra attention to remedy past inequities. This would be consistent with some notions of corrective justice. For others, success may come by simply eliminating any existing disparities. That is, all that is necessary is equality of treatment. In the language of this chapter, the first perspective would judge policy success as the presence of increased positive disparities in EPA and state enforcement of the CAA for environmental justice communities, whereas the second would characterize success as the reduction or elimination of disparities. One could argue for either characterization, and there is likely significant variance among stakeholders in the environmental justice advocacy community as to what a successful program's goals ought to include.

Depending on one's view of success, therefore, one could interpret our results in different ways. If success equates to the EPA and state agencies dedicating disproportionate enforcement toward facilities located in poor and minority communities, then EO 12898, the EPA's environmental justice strategy, and the other reforms of the mid-1990s largely failed. In only one of the twelve tests was EPA's policy associated with a post-policy change toward positive disparities. If, however, the goal is to erase disparities, our results suggest a more optimistic picture. In two-thirds of the cases, during the post-policy period, EPA or state enforcement rates were similar, regardless of the community's demographics. Both sets of agencies treated facilities located in poor and minority communities similar to those in nonpoor and nonminority communities, all else equal. It is important to note, however, that in just one of these cases did this reflect a movement from a pattern of negative disparities in the pre-policy period.

Although these conclusions clearly emerge from our data analysis, it is important to recognize several limitations of our study. First, we have analyzed only one federal pollution control law (albeit an important one). EPA and state enforcement responsibilities extend to many other areas of pollution control, ranging from water to hazardous waste. Without direct investigation of these and other areas of policy, we cannot reach conclusions about the overall policy effects on regulatory enforcement. Moreover, the analysis studied enforcement activity only through 2008, so it does not account for changes (positive or negative) in more recent years. The findings regarding state agencies, however, are generally consistent with a previous analysis of enforcement of the CWA and the RCRA, which also found little evidence that states gave more attention to facilities in poor and minority communities (Konisky 2009b).

Second, we have evaluated enforcement activity solely on the basis of the amount (or, more precisely, the probability) of an enforcement action being directed at regulated facilities. One could reasonably argue that the amount of enforcement activity may not reflect an efficient use of government enforcement resources when it comes to mitigating risks posed to society overall, or even for environmental justice target communities. However, while we do not contend that *more* enforcement activity necessarily means *more effective* enforcement, regulatory enforcement activity in general has been found to be an important determinant of pollution prevention and firm compliance (Gray and Shimshack 2011).

In reflecting on these results, one area that we would like to highlight is the findings regarding state enforcement. As noted previously, on balance, our analysis of CAA enforcement suggests that the federal reforms did not do much to push states toward more aggressive enforcement of major air polluters in target communities, with the exception of facilities in areas with large Hispanic populations. It is necessary to emphasize that the effect that we identified is the average effect across the fifty states, and there is undoubtedly large variance across the states, with some states acting as leaders and others as laggards. Identifying which states fall in which category is important for determining whether the outcomes were equally distributed across the country.

More generally, it is important to note that the EPA has not placed much emphasis in its most recent environmental justice initiative—*Plan EJ 2014*—on the hugely important role of state agencies in environmental enforcement. In fact, states are barely mentioned in the EPA's implementation plan for addressing equity issues through compliance and enforcement (EPA 2011). In our view, this is a glaring weakness in the new initiative, and the EPA should think carefully about how to use its financial influence and oversight powers to affect state enforcement behavior, as well as to further direct its attention to environmental justice priorities in performance partnership agreements and other federal-state cooperative arrangements.

Given the existence of varying perspectives of success and informational levels on EPA and state program activity, the implementation of any policy reform represents only one piece of an effective response to overburdened communities. The EPA's *Plan EJ 2014* highlights the essential role of communicating with communities, and we strongly echo this focus. It is important that EPA and state administrative agencies communicate their enforcement accomplishments with the affected communities. A successful change in enforcement activity may very well go unnoticed in a community that may traditionally be less engaged in politics or policy in a meaningful way. In the absence of effective communication, citizens' perceptions of an agency's continuing neglect is likely to become their reality. In this way, perceptions of environmental injustice may persist, despite the actions of government intended to genuinely address—though not necessarily resolve—disparities in regulatory enforcement.

Regulatory enforcement is a critical component of an effective environmental protection system. Laws matter only to the extent to which they are adhered, and our past work has shown some tendency for compliance rates, at least under the CAA, to be lower in some poor and minority communities (Konisky and Reenock 2013a, b). Moreover, enforcement represents a set of tools that government agencies can potentially use to remedy (or at least reduce) observed environmental risk disparities. Although the EPA and state governments cannot force power plants, factories, and other polluting facilities to reduce their emissions in low-income and minority communities (to the extent to which they are legally permissible under the law), they can ratchet up their enforcement efforts against firms that exceed emission limits. Thorough enforcement, performed fairly across all communities, is a goal that we believe EPA and state governments should work diligently to achieve.

Appendix

The appendix includes two tables pertaining to the data analysis in the chapter. Table A7.1 provides basic descriptive statistics for the variables included in the statistical analysis, and identifies the data source for each. Table A7.2 displays the results for the statistical models we discuss in the chapter.

Table A7.1
Descriptive Statistics

Variable	Data Source	Mean	Std. Dev.	Min	Max
Enforcement Measure					
EPA inspection	EPA's Integrated Data for Enforcement Analysis database	0.088	0.284	0.000	1.000
State inspection	EPA's Integrated Data for Enforcement Analysis database	0.537	0.499	0.000	1.000
EPA enforcement action	EPA's Integrated Data for Enforcement Analysis database	0.009	0.096	0.000	1.000
State enforcement action	EPA's Integrated Data for Enforcement Analysis database	0.090	0.286	0.000	1.000
EJ Indicators					
African-American target community	EPA's Integrated Data for Enforcement Analysis database	0.099	0.299	0.000	1.000
Hispanic target community	EPA's Integrated Data for Enforcement Analysis database	0.099	0.299	0.000	1.000
Poor target community	EPA's Integrated Data for Enforcement Analysis database	0.099	0.299	0.000	1.000
Demographic Indicators					
% African-American	U.S. Census Bureau	11.862	18.258	0.000	100.000
% Hispanic	U.S. Census Bureau	10.110	16.268	0.000	97.952
% Poverty	U.S. Census Bureau	14.507	8.926	0.000	92.177
% Household education	U.S. Census Bureau	55.569	15.865	0.000	100.000
Median household income (thousands)	U.S. Census Bureau	38.616	14.931	0.000	215.727

Table A7.1 (continued)

Policy Task Factor Indicators

Nonattainment	EPA's Green Book	0.387	0.487	0.000	1.000
Unemployment	Bureau of Labor Statistics	6.067	2.649	0.400	40.800

Political Context Indicators

Democratic governor	Calculated from original data	0.448	0.494	0.000	1.000
% Democrats in state Legislature	Calculated from original data	53.268	12.119	11.428	90.789
State government Ideology (ADA/ COPE)	Berry et al. (1998)	48.109	24.760	0.000	97.917

Firm-Level Indicators

EPA inspection (1 year lag)	EPA's Integrated Data for Enforcement Analysis database	0.088	0.284	0.000	1.000
State inspection (1 year lag)	EPA's Integrated Data for Enforcement Analysis database	0.537	0.499	0.000	1.000
EPA enforcement action (1 year lag)	EPA's Integrated Data for Enforcement Analysis database	0.009	0.096	0.000	1.000
State enforcement action (1 year lag)	EPA's Integrated Data for Enforcement Analysis database	0.090	0.286	0.000	1.000
Manufacturing firm	EPA's Integrated Data for Enforcement Analysis database	0.527	0.499	0.000	1.000
Utility firm	EPA's Integrated Data for Enforcement Analysis database	0.300	0.458	0.000	1.000
Oil and gas firm	EPA's Integrated Data for Enforcement Analysis database	0.052	0.223	0.000	1.000
Mining firm	EPA's Integrated Data for Enforcement Analysis database	0.011	0.104	0.000	1.000

Table A7.2
The Impact of EJ Policy Reform on EPA and State Enforcement Actions

	Model 1 — Logit, with Cubic Splines (EPA)				Model 2 — Logit, with Cubic Splines (States)			
	Pr(Inspection) b	s.e. b	Pr(Punitive Action) b	s.e. b	Pr(Inspection) b	s.e. b	Pr(Punitive Action) b	s.e. b
EJ Policy Indicators								
African-American target community	0.8579 **	0.1693	0.0552	0.2142	0.0945	0.0586	0.1797 *	0.0868
Hispanic target community	0.2976	0.2512	-0.2867	0.3257	-0.1810 **	0.0663	-0.9994 **	0.1368
Poor target community	-0.0602	0.1917	0.1708	0.1904	-0.0077	0.0533	0.0155	0.0893
EJ Policy 1995	1.4090 **	0.0877	0.5637 **	0.0998	0.5698 **	0.0216	0.3763 **	0.0387
AATC X Policy	-0.6949 **	0.1641	-0.0595	0.1884	-0.1642 **	0.0520	-0.1797 *	0.0789
HTC X Policy	-0.4368	0.2421	0.5341	0.3098	0.3855 **	0.0602	0.9873 **	0.1278
PTC X Policy	0.0539	0.1876	-0.1406	0.1941	-0.1505 **	0.0533	0.0307	0.0858
Demographic Controls								
Median household Income	-0.0002	0.0009	-0.0044	0.0026	0.0033 **	0.0008	-0.0004	0.0009
% African-American	-0.0041 **	0.0011	-0.0023	0.0026	0.0010	0.0008	-0.0002	0.0010
% Hispanic	-0.0009	0.0013	-0.0098 **	0.0034	-0.0049 **	0.0009	-0.0009	0.0011
% Household education	-0.0275 **	0.0010	-0.0071 **	0.0025	0.0053 **	0.0006	-0.0083 **	0.0007
% Poverty	-0.0066 **	0.0022	0.0060	0.0052	0.0115 **	0.0017	-0.0024	0.0021

Table A7.2 (continued)

Policy Task Controls										
% Unemployment	0.0449	**	0.0052	0.0202		0.0118	0.0108	**	0.0032	0.0233	**	0.0045
Nonattainment	0.0174		0.0217	0.3038	**	0.0556	0.0310		0.0178	0.2395	**	0.0219
Political Controls										
Democratic governor	1.0406	**	0.0342	−0.2700	**	0.0831	−0.0731	**	0.0228	−0.1586	**	0.0322
% Democrats in state legislature	0.0109	**	0.0013	0.0056		0.0031	0.0003		0.0009	−0.0103	**	0.0011
State government ideology	−0.0150	**	0.0008	0.0035		0.0019	0.0007		0.0005	0.0076	**	0.0007
Bush 1 administration dummy	−2.7147	**	0.1022	−1.3115	**	0.1180	−1.7990	**	0.0245	−1.2027	**	0.0461
Clinton administration dummy	−1.1504	**	0.0445	−0.0164		0.0774	−0.7431	**	0.0173	−0.2503	**	0.0238
Firm-Level Controls										
Utility firm	0.1792	**	0.0321	−0.0696		0.0985	0.0036		0.0242	−0.1511	**	0.0337
Manufacturing firm	0.1144	**	0.0311	0.4149	**	0.0886	0.1364	**	0.0231	0.3684	**	0.0285
Mining firm	−0.1456		0.0897	−0.0056		0.2382	0.3214	**	0.0768	0.3631	**	0.0916
Oil and gas firm	−0.2019	**	0.0620	−0.1266		0.2007	−0.1265	**	0.0395	0.1627	**	0.0520
State enforcement action (1 year lag)	0.2664	**	0.0253	1.0812	**	0.0619	0.4862	**	0.0252	0.7538	**	0.0218
Federal enforcement action (1 year lag)	0.4663	**	0.0679	0.7375	**	0.0687	0.5631	**	0.0683	−0.1499	**	0.0253
Intercept	0.5563	**	0.1446	−4.0295	**	0.3165	1.6624	**	0.0845	0.1587		0.5526
Log-likelihood	−48,435.13			−12,011.807			−10,6091.38			−64,353.61		
Cases	241,016			241,016			241,016			241,016		

Note: Cubic splines and EPA dummy variables not shown. *p < 0.05, **p < 0.01, two-tailed tests.

Notes

1. Until the CPI/NPR study, the contents of the watch list had not been made public. In response to this report, the EPA has since published a monthly update of the watch list on its website, although the specific criteria that lead to the inclusion or exclusion on the list are not made public (available at: http://www.epa-echo.gov/echo/echo_watch_list.html).

2. These differences are based on 2010 census data, measured at the ZIP code level.

3. The model also includes a spell-identification counter and three cubic spline variables (included to account for duration dependence) and an error term that allows for correlation of errors at the facility level.

4. More formally, we estimate the following model: $E_{it} = \beta_0 + \beta_1 EJ_{it} + \beta_2 Policy_t + \beta_3 EJ_{it}{}^* Policy_t + \beta K Controls_{it} + \beta S Splines_{it} + \varepsilon_{it}$, where i indexes facilities and t indexes years; E is a measure of EPA or state regulatory enforcement of the CAA; EJ is a vector of demographic characteristics that identifies environmental justice communities; $Policy$ is an indicator variable capturing the period before and after the environmental justice policies were put in place; $Controls$ is a vector of k control variables to capture other correlates of regulatory enforcement outcomes; $Splines$ is a vector that includes a spell-identification counter, as well as three cubic spline variables (included to account for duration dependence); and ε is an error term that allows for correlation of errors at the facility level. The effect of the policy intervention is given by the following partial derivative: $\partial E_{it}/\partial Policy_t = \beta_1 + \beta_3 EJ_{it}$.

5. We elected to focus on currently active major air sources because of the availability of geospatial information that enables us to precisely locate them in geographical space. Geospatial data, however, are consistently available only for major air sources that are currently active, so we restricted the data to these facilities. We also make the assumption that these facilities were active for the entire period of study, which is a necessary assumption because the EPA does not track the historical operation status (active or inactive) of regulated entities.

6. Because of missing data in some decennial Census years, the number of facilities included in the analysis varies by a small number in the 1990, 2000, and 2010 information.

7. Most studies in the environmental justice literature use Census units such as counties, ZIP codes, or tracts to develop demographic measures of the relevant community, but there are two well-known problems with this approach. First, demographic characteristics may not be uniformly distributed among the population of the Census unit, a particular problem for geographically larger units. This assumption is maintained with the areal apportionment method, but it is less problematic because Census tracts are relatively small geographical units. Second, the relevant population may extend to adjacent Census units, especially when a facility is located on the border of another unit.

8. Specifically, we separately intersect a geospatial map of the major air sources with a geospatial map of U.S. Census tracts. Using a predefined distance of 1 mile to indicate the potentially affected population, we create a 1-mile radius around each facility as a buffer and intersect the maps to merge spatial data from the circular buffer with that from the Census maps. The intersections are then used

as weights for each demographic attribute, where the weight is the proportion of each Census unit contained within the circular buffers. We then create a weighted average of the relevant demographic data to create a measure for each variable of interest within 1-mile of the facility (i.e, percent African Americans, percent Hispanic, percent poor, percent with at least a high school education, and median household income). In a small number of cases, the demographic data from the Census was missing for some Census tracts; in these cases, we imputed a value based on the average of the other intersected units.

9. We also repeated all of our analysis using different distances (specifically, half-mile and 2-mile radii) to define the neighborhood around facilities. The half-mile analysis yielded substantively nearly identical results (the interaction with African-American communities is null for the state punitive action model). The 2-mile analysis generated two additional differences. With the expanded 2-mile radii, we also find that the estimated effect for the policy in poor communities with respect to state inspections is now null, and that the previously estimated null effect for the policy in African-American communities is now statistically significant for state punitive actions.

10. Other researchers have used a less-restrictive threshold based on the top quartile of an area (e.g., Lavelle and Coyle 1992; Ringquist 1998). In work not reported here, we conducted an analysis using this alternative threshold, with substantively similar results.

11. The thresholds for our other areal boundaries (half-mile and 2-mile) were 40 percent, 32 percent, and 28 percent, and 37 percent, 30 percent, and 25 percent for African-American, Hispanic, and poor communities, respectively. For the half-mile boundary, within these communities, the population demographic statistics for each targeted group were mean = 60 percent and standard deviation (s.d.) = 16 percent; mean = 54 and s.d. = 17 percent; and mean = 36 percent and s.d. = 8 percent, respectively. For the 2-mile boundary, within these communities, the population demographic statistics for each targeted group were mean = 54 percent and s.d. = 14 percent; mean = 50 percent and s.d. = 16 percent; mean = 31 percent and s.d. = 5.5 percent, respectively.

12. We also coded the intervention variable split on 1994, which produced similar findings. In addition, we considered a balanced policy window where we limited our analysis to the pre-reform era (1990–1994) and the post-reform era (1995–1999). This truncated data set (1990–1999) yielded similar findings to those reported in this chapter.

13. For each of the models, log-likelihood tests support the necessity of including cubic splines for each of the dependent variables to account for time dependence between observations for a given individual firm. These tests suggest negative duration dependence for each dependent variable, with the hazard rate of either an inspection or a punitive action declining in its duration.

14. A few of our estimated effects for our control variables are not in line with expectations from the literature. Democratic control of either the governor's office or the state legislature and the ideological liberalism of state government is generally thought to increase the probability of firm inspection and punishment—but in a few instances, we have found the opposite. The effect of a Democratic governor is the most consistent of these unexpected findings.

References

Berry, William D., Evan J. Ringquist, Richard C. Fording, and Russell L. Hanson. 1998. Measuring Citizen and Government Ideology in the American States, 1960–93. *American Journal of Political Science* 42: 327–348.

Browner, Carol M. 1995. *The EPA's Environmental Justice Strategy.* Washington, DC: Environmental Protection Agency.

Bryant, Bunyan. 1995. *Environmental Justice: Issues, Policies, and Solutions.* Washington, DC: Island Press.

Bullard, Robert D. 1993. Anatomy of Environmental Racism and the Environmental Justice Movement. In *Confronting Environmental Racism: Voices from the Grassroots,* ed. Robert D. Bullard, 15–39. Boston: South End Press.

Bullard, Robert D., and Glenn S. Johnson. 2000. Environmental Justice: Grassroots Activism and Its Impact on Public Policy Decision Making. *Journal of Social Issues* 56 (3): 555–578.

Center for Public Integrity and National Public Radio (CPI/NPR). 2011. *Poisoned Places: Toxic Air, Neglected Places.* http://www.publicintegrity.org/environment/pollution/poisoned-places.

Collin, Robert W. 1993. Environmental Equity and the Need for Government Intervention: Two Proposals. *Environment* 35: 41–43.

Earnhart, Dietrich. 2004a. Regulatory Factors Shaping Environmental Performance at Publicly Owned Treatment Plants. *Journal of Environmental Economics and Management* 48: 655–681.

Earnhart, Dietrich. 2004b. The Effects of Community Characteristics on Polluter Compliance Levels. *Land Economics* 80: 408–432.

Environmental Protection Agency (EPA). 1993. *1993 Environmental Justice Initiative.* Washington, DC: Environmental Protection Agency.

Environmental Protection Agency (EPA). 1998. Issuance of Policy on Timely and Appropriate Enforcement Response to High Priority Violations. Memorandum from Eric Schaeffer. Washington, DC: Office of Enforcement and Compliance Assurance. December 22.

Environmental Protection Agency (EPA). 1999. The Timely and Appropriate Enforcement Response to High-Priority Violations. Memorandum from Eric Schaeffer. Washington, DC: Office of Enforcement and Compliance Assurance. June 23.

Environmental Protection Agency (EPA). 2008. *Reevaluation of the Use of Recidivism Rate Measures for EPA's Civil Enforcement Program.* Washington, DC: Office of Enforcement and Compliance Assurance.

Environmental Protection Agency (EPA). 2011. Advancing Environmental Justice through Compliance and Enforcement: Implementation Plan. Washington, DC: Environmental Protection Agency (EPA). September. http://www.epa.gov/environmentaljustice/plan-ej/ce-initiatives.html.

Gray, Wayne B., and Ronald J. Shadbegian. 2004. "Optimal" Pollution Abatement: Whose Benefits Matter, and How Much? *Journal of Environmental Economics and Management* 47: 510–534.

Gray, Wayne B., and Ronald J. Shadbegian. 2012. Spatial Patterns in Regulatory Enforcement: Local Tests of Environmental Justice. In *The Political Economy of Environmental Justice*, ed. H. Spencer Banzhaf, 225–248. Palo Alto: CA: Stanford University Press.

Gray, Wayne B., and Jay P. Shimshack. 2011. The Effectiveness of Environmental Monitoring and Enforcement: A Review of the Empirical Evidence. *Review of Environmental Economics and Policy* 5 (1): 3–24.

Hamilton, James T. 1995. Testing for Environmental Racism: Prejudice, Profits, Political Power? *Journal of Policy Analysis and Management* 14 (1): 107–132.

Hamilton, James T., and W. Kip Viscusi. 1999. *Calculating Risks? The Spatial and Political Dimensions of Hazardous Waste Policy.* Cambridge, MA: MIT Press.

Hastings College of the Law. 2004. *Environmental Justice for All: A Fifty-State Survey of Legislation, Policies, and Initiatives.* January 2004. Public Law Research Institute, University of California.

Helland, Eric. 1998. The Enforcement of Pollution Control Laws: Inspections, Violations, and Self-Reporting. *Review of Economics and Statistics* 80 (1): 141–153.

Konisky, David M. 2007. Regulatory Competition and Environmental Enforcement: Is There a Race to the Bottom? *American Journal of Political Science* 51 (4): 853–872.

Konisky, David M. 2009a. Inequities in Enforcement? Environmental Justice and Government Performance. *Journal of Policy Analysis and Management* 28 (1): 102–121.

Konisky, David M. 2009b. The Limited Effects of Federal Environmental Justice Policy on State Enforcement. *Policy Studies Journal: The Journal of the Policy Studies Organization* 37 (3): 475–496.

Konisky, David M., and Christopher Reenock. 2013a. Case Selection in Public Management Research: Problems and Solutions. *Journal of Public Administration: Research and Theory* 23 (2): 361–393.

Konisky, David M., and Christopher Reenock. 2013b. Compliance Bias and Environmental (In)justice. *Journal of Politics* 75 (2): 506–519.

Konisky, David M., and Tyler S. Schario. 2010. Examining Environmental Justice in Facility-Level Regulatory Enforcement. *Social Science Quarterly* 91 (3): 835–855.

Kuehn, Robert R. 2000. A Taxonomy of Environmental Justice. *Environmental Law Reporter* 30: 10681–10703.

Lavelle, Marianne, and Marcia Coyle. 1992. Unequal Protection: The Racial Divide in Environmental Law. *National Law Journal* (September): 21.

Leighley, Jan E., and Arnold Vedlitz. 1999. Race, Ethnicity, and Political Participation: Competing Models and Contrasting Explanations. *Journal of Politics* 61 (4): 1092–1114.

Lester, James P., David W. Allen, and Kelly M. Hill. 2001. *Environmental Injustice in the United States: Myths and Realities.* Boulder, CO: Westview Press.

Mohai, Paul, and Robin Saha. 2006. Reassessing Racial and Socioeconomic Disparities in Environmental Justice Research. *Demography* 43 (2): 383–399.

Mohai, Paul, and Robin Saha. 2007. Racial Inequality in the Distribution of Hazardous Waste: A National-Level Reassessment. *Social Problems* 54 (3): 343–370.

National Environmental Justice Advisory Council (NEJAC). 1995. *Achieving Environmental Protection: Compliance, Enforcement, and Environmental Justice.* Enforcement Subcommittee, Environmental Protection Agency.

Rechtschaffen, Clifford, and David L. Markell. 2003. *Reinventing Environmental Enforcement & The State/Federal Relationship.* Washington, DC: Environmental Law Institute.

Ringquist, Evan J. 1998. A Question of Justice: Equity in Environmental Litigation, 1974–1991. *Journal of Politics* 60 (4): 1148–1165.

Ringquist, Evan J. 2005. Assessing Evidence of Environmental Inequities: A Meta-Analysis. *Journal of Policy Analysis and Management* 24 (2): 223–247.

Ringquist, Evan J., and David H. Clark. 1999. Local Risks, States' Rights, and Federal Mandates: Remedying Environmental Inequities in the U.S. Federal System. *Publius: The Journal of Federalism* 29 (2): 73–93.

Rosenstone, Steven J., and John Mark Hansen. 1993. *Mobilization, Participation, and Democracy in America.* New York: Macmillan.

Shadish, William R., Thomas D. Cook, and Donald T. Campbell. 2002. *Experimental and Quasi-Experimental Designs for Generalized Causal Inference.* Boston: Houghton Mifflin.

Scholz, John T., and Cheng-Lung Wang. 2006. Cooptation or Transformation? Local Policy Networks and Federal Regulatory Enforcement. *American Journal of Political Science* 50 (1): 81–97.

Shogren, Elizabeth, Kristen Lombard, and Sandra Bartlett. 2011. Where Regulators Failed, Citizens Took Action—Testing Their Own Air, November 10. http://www.publicintegrity.org/2011/11/10/7355/where-regulators-failed-citizens-took-action-testing-their-own-air.

Sigman, Hilary. 2003. Letting States Do the Dirty Work: State Responsibility for Federal Environmental Regulation. *National Tax Journal* 56 (1): 107–122.

Verba, Sidney, and Norman H. Nie. 1972. *Participation in America: Political Democracy and Social Equality.* New York: Harper and Row.

Verba, Sidney, Kay Lehman Schlozman, and Henry Brady. 1995. *Voice and Equality: Civic Voluntarism in American Politics.* Cambridge, MA: Harvard University Press.

8

Environmental Justice in the Courts

Elizabeth Gross and Paul Stretesky

The environmental justice movement recognizes the importance of social equity when assessing environmental burdens and benefits (Bullard 1990; Gould, Pellow, and Schnaiberg 2008). As a result, environmental justice concerns are diverse and focus on a variety of issues. As Bunyan Bryant (1995, 6) makes clear, environmental justice "refers to those cultural norms and values, rules, regulations, behaviors, policies, and decisions to support sustainable communities where people can interact with confidence that the environment is safe, nurturing, and productive [and is] is supported by decent paying safe jobs, quality school and recreation; decent housing and adequate health care; democratic decision-making and personal empowerment."

The focus of this chapter is on environmental justice in the courts, which makes up only one small portion of the movement's overall concerns and activities. More specifically, this chapter emphasizes corrective justice, or the role of contemporary U.S. courts in punishing environmental offenders, reversing environmental injustices, or both (Weinrib 2012). Prior to examining the way courts function with respect to environmental justice, we briefly examine major federal environmental justice policies that attempt to direct environmental enforcement. We focus on federal environmental justice policy because this is the level of governance at which a significant amount of court activity has occurred. We draw upon that policy to examine whether federal agencies have pursued environmental justice goals set out in those policies in the courts.

President Bill Clinton's executive order on environmental justice did not stress the role of the courts as a way to address environmental equity. In fact, the only mention of judicial review in Executive Order 12898 on Environmental Justice (EO 12898) highlighted its limits. Nevertheless, evaluating environmental justice in the courts is important to consider for a couple of reasons. First, the courts are often the end point of enforcement cases brought by the Environmental Protection Agency (EPA) and other agencies against regulated entities for violating environmental laws. Their decisions, therefore, are a critical component of enforcement. Second, many in the environmental justice scholarly and advocacy communities have argued that the courts provide an important venue for holding government accountable under existing statutes. This was reflected in Clinton's memorandum that accompanied EO 12898, which emphasized that Title VI of the Civil Rights Act of 1964 and the National Environmental Policy Act were important vehicles through which environmental justice claims might be addressed.

For these reasons, consideration of environmental justice in the courts is an important component in any evaluation of federal environmental justice policy. With this in mind, the first part of the chapter examines if the courts encourage environmental injustice by providing lenient sentences to some environmental offenders. Overall, the research on this issue suggests that the monetary fines and penalties that are handed down from the courts are not directly discriminatory—that is they do not appear to disproportionately penalize industries in white and affluent communities and ignore violations in poor and minority communities. Instead, it appears that if discrimination in sentencing exists it is likely to be indirect.

In the next section, we question whether the courts are correcting environmental injustice by ruling that the unequal siting of facilities violates civil rights law. We look to see if courts, as potential agents of positive change, diminish environmental injustice by opening up opportunities for communities to challenge discriminatory siting. We carry out this investigation in the case of Title VI of the federal Civil Rights Act, the Equal Protection Clause of the Fourteenth Amendment of the U.S. Constitution, and the National Environmental Policy Act (NEPA). Unfortunately, we discover that although the EPA continually emphasizes the need for effective implementation of Title VI, the courts appear hesitant to use the Civil Rights Act to achieve environmental justice. This is most apparent in the U.S. Supreme Court decision *Alexander v. Sandoval* (2001), which, while not an environmental justice case itself, does operate to significantly limit the effectiveness of the Civil Rights Act for environmental justice plaintiffs. We suggest that the cases after *Alexander v.*

Sandoval demonstrate the lack of commitment within courts to support environmental justice efforts. We, unfortunately, discover a similar lack of commitment with regard to the Equal Protection Clause and NEPA in environmental justice cases. Finally, we conclude this chapter by proposing potential studies and strategies for addressing environmental injustice and achieving favorable environmental justice outcomes.

Federal Environmental Justice Policy and the Role of the Courts

The role of courts in environmental policy has a long history in the United States (McSpadden 1995; Melnick 1983; O'Leary 1993). Courts may adjudicate environmental violations, shape the definition of environmental problems through their interpretations of environmental law, decide whether environmental laws are constitutional, determine which agencies have authority over environmental management issues, and engage in judicial review of agency decisions (Burns, Lynch, and Stretesky 2009). As a result, courts influence environmental justice outcomes through cases brought before them for adjudication, appeal, and review. Because agency priorities may influence the types of cases that are brought before the courts, we highlight the priorities identified in EO 12898, the EPA's recently developed environmental justice strategy, *Plan EJ 2014* (EPA 2013), as well as the role of Title VI.

Among the most important environmental justice policy developments in the United States was the issuance of EO 12898 on February 11, 1994, by President Clinton (Bowers 1995; Bullard and Johnson 2000; O'Neil 2007). Section 1–1 of EO 12898 states, "Each Federal agency shall make achieving environmental justice part of its mission by identifying and addressing, as appropriate, disproportionately high and adverse human health or environmental effects of its programs, policies, and activities." Importantly, EO 12898 applies to all federal agencies, two of which are especially important to the order's ideals: the EPA and the Department of Justice (DOJ). The EPA issues pollution permits, may bring administrative and civil actions against offenders that violate environmental regulations, and participates in the mitigation of uncontrolled and abandoned waste sites throughout the country, a practice that often involves courts. The DOJ may initiate criminal cases against offenders who violate environmental laws (Burns, Lynch, and Stretesky 2009). Unfortunately, research on the effectiveness of EO 12898 appears to suggest that the policy has had little impact on environmental justice outcomes in either agency (Konisky 2009a; Murphy-Green and Leip 2002; O'Neil 2007; Simms 2013). Bonner (2012, 100) perceives, "while the objectives of EO 12898 aim to mitigate the adverse impact of environmental policies on minorities, it is unclear whether it has been successful." It is also important to point out that EO 12898 is not directly enforceable in the courts.

As discussed next, EO 12898 can potentially influence the EPA and DOJ through its implementation and use of Title VI of the Civil Rights Act of 1964. This point was emphasized in Clinton's memo that accompanied the order (Clinton 1994). Specifically, the president states:

In accordance with Title VI of the Civil Rights Act of 1964, each Federal agency shall ensure that all programs or activities receiving Federal financial assistance that affect human health or the environment do not directly, or through contractual or other arrangements, use criteria, methods, or practices that discriminate on the basis of race, color, or national origin.

Title VI aims to stop discrimination by entities that receive federal funds. In theory, the EPA can use Title VI to prevent the federal funding of state and local governments that issue pollution permits that create a disparate impact. Thus, Title VI has been viewed as a powerful weapon in fighting environmental injustice, especially given the priorities in EO 12898. According to the EPA, however, there are relatively few Title IV investigations and even fewer Title VI settlements (http://www.epa.gov/civilrights/TitleVIcases/index.html). The DOJ has not brought any Title VI cases forward to date. The lack of Title VI cases suggests that the courts have played a very limited role in achieving environmental justice though Title VI. As discussed later in this chapter, the court has also interpreted Title VI narrowly, especially in relation to private actions, and this interpretation has limited its effectiveness and therefore conflicts with the spirit of EO 12898 as Clinton (1994) first envisioned it.

When EO 12898 was established, EPA administrator Carol M. Browner (1995, 13) reported in the agency's *Environmental Justice Strategy* that "strong and effective enforcement of environmental and civil rights laws is fundamental to virtually every mission of EPA." Recently, environmental justice organizations have renewed their attempts to advocate for civil rights through Title VI. As a result, federal agencies appear to have a renewed interest in achieving environmental justice outcomes.

For example, the EPA recently reconvened the Federal Interagency Working Group on Environmental Justice in 2010 for the first time in over a decade. The EPA (2013, 1) states in its *Plan EJ 2014 Progress Report* that "for the first time in its 42-year history, the U.S. Environmental Protection Agency (EPA) has laid the cornerstones for fully implementing its environmental justice (EJ) mission of ensuring environmental protection for all Americans, regardless of race, color, national origin, income, or education." *Plan EJ 2014* includes a new emphasis on Title VI, and the EPA (2011, 31) states that "administrator Lisa Jackson has made a commitment to reform and revitalize the Agency's Title VI program, which includes ensuring that recipients of EPA Federal Assistance comply with their Civil Rights requirements."

In addition to the focus on using Title VI, the Council on Environmental Quality has used EO 12898 to guide NEPA assessments and environmental impact statements that describe the effects of federal agency actions. The idea is that NEPA must consider environmental justice issues when preparing impact statements. When potential environmental injustice is identified, proposed mitigating efforts should be considered. Unfortunately, the federal government is not required to adopt any environmental justice recommendations within NEPA. Because agencies need only consider environmental impacts, EO 12898 may be prevented from being a more effective policy tool for achieving environmental justice (Bratspies et al. 2008). As we suggest below, NEPA is still not used to achieve environmental justice. Nevertheless, the EPA (2011, 32) has recently suggested that NEPA will be used much more aggressively in the future.

The Distribution of Court Penalties

The environmental justice movement has a long history in the United States and can be interpreted as emerging from a series of complex historical events that have brought together civil rights activists, traditional environmentalists, anti-toxic activists, workers, and academics (Taylor 2000; Cole and Foster 2001). Within this movement, these groups of actors share common "motives, background, and perspectives" (Cole and Foster 2001, 32). In particular, academics are one set of actors producing studies that are used in the courts to demonstrate how environmental hazards are distributed by race and class (e.g., Bullard 1983; Mennis 2005). For example, Robert Bullard's work on waste incinerators in Houston, Texas, is seen as critical to the movement's academic roots (Bullard 1990; Cole and Foster 2001). Bullard's early efforts at bringing empirical evidence of environmental inequity directly to the courts has generated considerable attention from the movement, as these studies provide an opportunity to help frame grievances (see Cole and Foster 2001, 25).

The first academic study to examine whether courts are biased in their application of penalties did not occur until 1992, when Lavelle and Coyle published an article that examined whether race and class inequalities exist in the distribution of monetary penalties imposed by the courts for violations of the Clean Air Act (CAA); the Clean Water Act (CWA); the Safe Drinking Water Act (SDWA); the Comprehensive Environmental Response, Compensation, and Liability Act (CERCLA); and the Resource Conservation and Recovery Act (RCRA). Importantly, their study examined only the distribution of penalties among all federal violations contained in the EPA Administrative Enforcement Dockets database, which lists administrative and civil violations between the years of 1985 and 1991. While their results are limited to noncriminal actions in federal courts, the implications for environmental justice are significant. As Ringquist (1998, 1151) notes, "The study received a good deal of press coverage [and] is often cited by environmental justice advocates as an example of how government actions discriminate against minorities and the poor to perpetuate inequalities in exposure to pollution."

Lavelle and Coyle (1992) studied injustice in the courts by obtaining the racial and economic demographics of each ZIP code where at least one environmental violation occurred. ZIP codes were ranked by race and income and then compared according to the mean monetary penalties reported in the Docket database. The researchers found that violations in white ZIP codes received penalties that averaged $153,607, while violations that occurred in minority ZIP codes received penalties that averaged $105,028. They also discovered that the mean penalty in high-income ZIP codes was higher ($146,993) than the mean penalty in low-income ZIP codes ($95,564). But these results did not hold up when the authors examined the type of violation. For example, CERCLA violations received *higher* fines in minority ZIP codes than in white ZIP codes. In addition, higher penalties were found in low-income ZIP codes than in high-income ZIP codes except for CWA violations. Lavelle and Coyle's results are suggestive of environmental injustice in the courts in the case of race, but their findings are far from definitive, especially in the case of income. Importantly, the researchers did not control for competing variables that may have produced the differences they observed.

Ringquist (1998) also examined penalties between the years 1974 and 1991 and found little evidence of environmental injustice in sentencing disparity after adjusting for case characteristics, judge attributes, and the political environment. Moreover, his analysis suggested that a 1 percent increase in the percentage of minorities living in a given ZIP code was associated with a 1 percent *increase* in an environmental fine. This relationship is the opposite of what the environmental injustice hypothesis would predict, since penalties in ZIP codes that are predominately African-American are larger than penalties in ZIP codes that are largely white.

One additional study by Atlas (2001) also examined the Docket database to see if there is significant discrimination against minorities and the poor in the case of civil penalties for environmental violations. Atlas (2001) examined Docket violations by Census tracts rather than ZIP codes and controlled for many of the same variables as Ringquist; and he discovered that minority neighborhoods were fined an average of $133,808, while white areas were fined an average of $113,791. This finding led him to conclude that there is little "basis for concluding that penalties are lower in disproportionately minority or low-income areas." Taken together, research on civil penalties suggests that courts do not discriminate by race or income when penalizing environmental violators.

These findings that there was no relationship between court decisions and punishment can be generalized to other types of court penalty studies as well. For example, Lynch, Stretesky, and Burns (2004) examined environmental fines leveled against 153 petroleum refineries for administrative, civil, and criminal violations of environmental laws. These environmental violations were primarily adjudicated in federal courts. The authors used race and economic demographics at both the Census tract and ZIP code levels to study the community characteristics around refineries that violate the law. When studying Census tracts, they discovered that after controlling for the type of crime and company characteristics, court-ordered fines were approximately equal for refineries operating in white and minority areas. This was not the case for ZIP codes. Cross-sectional results suggest that a standard deviation increase of 1 in the percentage of Hispanic residents was associated with a 95 percent decrease in the penalty amount. The researchers could not rule out the possibility of aggregation effects, but they noted that this result suggests that refineries situated in Hispanic ZIP codes are much less likely to be punished by the courts than refineries located in white ZIP codes.

Research has also been conducted on criminal violations. Greife (2012), for example, examined environmental justice issues in the case of penalties against companies found guilty for criminal violations of environmental law. Greife focused only on violations adjudicated by the DOJ and found that while crime seriousness and company characteristics were good predictors of monetary penalties, the average income of residents living within three miles of the violation location was also an important predictor. Specifically, a 1 percent increase in residents earning more than $75,000 in annual income was associated with an increase in criminal penalties on the order of $236,297. He did not find evidence of environmental injustice with respect to race. Greife also conducted a post hoc analysis, and he argues that prosecutors "overcharge" those environmental offenders who commit their crimes near wealthy residents. He demonstrates this argument empirically by showing that maximum allowable fines are elevated in high-income neighborhoods, and that this elevated potential is what leads the court to give higher-than-average penalties to offenders. Greife's results are interesting in that they are similar to those of Ringquist and Atlas. However, with respect to criminal violations, the economic injustice that may be occurring when punishing environmental offenders may be a function of prosecutors, not the court.

More research is needed to examine the behavior of investigative agencies and prosecuting attorneys when examining environmental penalties. As Simms (2013, 16) notes, "Environmental justice advocates have long protested … relative inattention of environmental enforcement officials to violations that primarily affect these communities." Specifically, researchers should examine where and how investigations are carried out and the willingness of prosecutors and regulatory attorneys to enforce environmental violations by filing those violations in the courts. Our review suggests that there is little discrimination in courts sentencing environmental offenders. This is not surprising, given that prosecutors and regulatory attorneys have considerably more discretion than judges because prosecutors and regulatory attorneys have the discretion to move a case forward. As a result, biases in sentences may not be appearing because prosecutors have brought forth only cases that are "worthy" of enforcement (Davis 1998). Missing from statistical analysis of sentencing disparity are cases that may not receive attention because of community characteristics. We therefore encourage future researchers to examine the potential role of investigators and prosecutors in detecting environmental violations and bringing them forward to the courts. Specifically, do extra legal case characteristics determine if cases are brought forward to prosecutors and the courts? Do community characteristics influence inspection levels or the severity of charges filed in court? Answering these questions is critical to understanding how environmental injustice develops and if courts can play a role in mitigating that injustice.

The Role of Courts in Achieving Environmental Justice

The general role of the courts in environmental justice–related claims is to consider the plaintiff's claim and determine whether a violation of law has been proved and, if so, what action will be taken to address the proved violation. Often, environmental justice claims are brought by citizens who sue the government on the grounds that it made a siting or permitting decision that has affected or will affect their community adversely (Cole 1993). The courts consider evidence and argument and, if a violation is established, could issue an injunction to prohibit the government's action from taking effect. Depending on the kind of case and claims proved, the court could order damages—at least in theory. In practice, however, it is difficult to prove a violation of law on environmental justice grounds; as a result, the courts have generally not issued orders against alleged environmental justice offenders.

In this section, we examine common legal claims asserted by plaintiffs in the courts. We examine the extent to which these lawsuits result in outcomes that may reduce environmental injustice in the case of Title VI, the Equal Protection Clause, and the NEPA.

Title VI of the 1964 Civil Rights Act

Environmental justice plaintiffs often utilize Title VI of the Civil Rights Act of 1964 to try to establish that they have been illegally subjected to disproportionate environmental consequences. A large body of legal scholarship has outlined various arguments for using Title VI to redress violations of environmental justice (e.g., Cole 1991; Godsil 1991; Hammer 1996; Mank 1999a, b, 2008). The EPA itself has issued guidance over the years to clarify the agency's interpretation and implementation of Title VI. In 1998, the EPA issued draft guidance establishing how the agency would investigate Title VI complaints and environmental justice challenges to permits (EPA 1998). This guidance, particularly the portion that outlined the disparate impact analysis that the agency would use to evaluate state-issued permits, was met with stiff resistance. State and local governments, represented by groups such as the Environmental Council of the States, the National Governors Association, the National Association of Counties, and the U.S. Conference of Mayors, complained that the guidance was too vague and infringed on their local land use authority. Government officials, joined by members of the business community, also objected on the grounds that the guidance would interfere with efforts to economically revitalize urban areas (Ringquist 2004). Members of Congress complained that the guidance would stifle economic development, and from 1998 to 2001, it enacted appropriations bills that suspended the EPA's authority to accept new Title VI cases. The EPA issued revised guidance in 2000 to address some of the previously raised concerns, but they remain unsatisfactory to many environmental justice stakeholders.

Important Case Law

The extent to which environmental justice claims can be redressed through Title VI is also highly dependent on case law. Two provisions of Title VI—Section 601 and Section 602—are most relevant. Section 601 of Title VI provides that no person shall, "on the ground of race, color, or national origin, be excluded from participation in, be denied the benefits of, or be subjected to discrimination under any program or activity receiving Federal financial assistance" (Civil Rights Act, 1964). But in order to prevail on a Section 601 claim, plaintiffs must prove intentional discrimination. For example, in *Alexander v. Sandoval* (2001, 275, 280), the U.S. Supreme Court stated that "it is ... beyond dispute ... that § 601 prohibits only intentional discrimination." However, it has been recognized that racial and ethnic discrimination is an "often-intractable problem" and that "it is often difficult to obtain direct evidence of [the] motivating animus" (306, 307n13 [dissent]). The observation that it is difficult to find direct evidence of discrimination is confirmed by scientific studies that find that racial and economic segregation around environmental hazards may intensify over time, in what can only be described as indirect discrimination (Stretesky and Hogan 1998). As a result, the court's focus on direct discrimination is unlikely to increase environmental equity. Specifically, it ignores larger social forces that may be discriminatory and therefore contribute to environmental injustice (Stretesky and Hogan 1998). In addition, it is well recognized that it is exceedingly hard to establish proof of direct discrimination (Davis 1998).

One example of the way Section 601 of Title VI may play out in courts is through cases such as *South Camden Citizens in Action v. New Jersey Department of Environmental Protection* (2006; hereafter referred to as *South Camden 2006*), a case which was first heard in the U.S. District Court for the District of New Jersey in 2001, was appealed to the Court of Appeals for the Third Circuit, and was then remanded back to the district court level for further proceedings. In that case, the New Jersey Department of Environmental Protection (NJDEP) issued a permit for the construction and operation of a grinding facility in a neighborhood where 91 percent of the residents were minority. South Camden Citizens argued intentional discrimination in the siting of the facility and introduced statistical evidence by Mennis (2005) to show a pattern of environmental injustice in New Jersey. While the court recognized that the environmental effects of the facility siting could bear more heavily on minorities than nonminorities, it based its logic on Mennis's own testimony to state that "any disparate impact that exists does not, by itself, support a finding of a discriminatory purpose on the part of NJDEP" and then concluded that "this impact alone is not determinative..." (*South Camden* 2006, 75–76). The court suggested in a footnote that South Camden Citizens "should direct their efforts prospectively to the appropriate legislative and agency forums and work towards a sensible and meaningful environmental equity policy for the future" (116).

Just as environmental justice plaintiffs have encountered difficulty in successfully prosecuting environmental justice claims under Section 601 of Title VI, they have also had difficulty using Section 602 of the Civil Rights Act to achieve justice. Section 602 of Title VI (Section 2000d-1) provides that:

Each Federal department and agency which is empowered to extend Federal financial assistance to any program or activity, by way of grant, loan, or contract other than a contract of insurance or guaranty, is authorized and directed to effectuate the provisions of section 2000d [Section 601] of this title with respect to such program or activity by issuing rules, regulations, or orders of general applicability which shall be consistent with achievement of the objectives of the statute authorizing financial assistance in connection with which the action is taken.

Section 602 provides the groundwork for the EPA to make regulations that prohibit agency funding recipients from engaging in actions that have a discriminatory effect. As discussed previously, the EPA has enacted regulations that attempt to prohibit recipients of EPA funding from administering programs that have a discriminatory effect. More specifically, the EPA regulations provide that:

A recipient shall not use criteria or methods of administering its program which have the effect of subjecting individuals to discrimination because of their race, color, national origin, or sex, or have the effect of defeating or substantially impairing accomplishment of the objectives of the program with respect to individuals of a particular race, color, national origin, or sex. (Nondiscrimination in Programs Receiving Federal Assistance from the Environmental Protection Agency 2012, 40 C.F.R. § 7.15).

The EPA's focus on "effect" rather than "intent" is promising. For instance, when the South Camden Citizens first took their case to the U.S. District Court for New Jersey in 2001, it looked as though it would be a victory for the environmental justice movement. The plaintiffs asserted that Section 602 provides a private cause of action under Title VI (*South Camden* D.N.J. 2001; hereafter called *South Camden 2001*). Specifically, the plaintiffs argued that Section 602's authorization of the EPA to issue the rule prohibiting discriminatory effects provided plaintiffs with the power to sue "on a theory of disparate impact discrimination in the administration of a federally funded program" (474). The court proceeded to consider the "novel question of whether a recipient of EPA funding has an obligation under Title VI to consider racially discriminatory disparate impacts when determining whether to issue a permit, in addition to compliance with applicable environmental standards" (474). In its analysis, the court initially recognized that "Section 602 of Title VI clearly authorizes federal agencies, such as the EPA, to promulgate regulations implementing Section 601" (475). Further, the court noted that the EPA has interpreted Title VI to prohibit EPA funding recipients from "utilizing 'criteria and methods' which have the 'purpose or effect' of discrimination against individuals based on their race, color, or national origin" [476, quoting 40 C.F.R. § 7.35(b)]. The court also found that these EPA regulations applied to permitting decisions

(*South Camden* 2001). Ultimately, the *South Camden* ruling granted the plaintiffs' request for a declaratory judgment that the state agency "violated Title VI of the Civil Rights Act by failing to consider the potential adverse, disparate impact of the ... facility's operation on individuals based on their race, color, or national origin, as part of its decision to permit [the] proposed facility" (481). The *South Camden* decision further granted the plaintiffs' request for a preliminary injunction, vacating the permits that New Jersey granted to the facility and enjoining the NJDEP from issuing permits to the company unless a protocol for reviewing permit applications to ensure compliance with Title VI was implemented (*South Camden* 2001, 481). The decision in *South Camden* to recognize the private right of action for environmental justice plaintiffs under Section 602, to recognize that environmental justice plaintiffs may prevail by showing a discriminatory impact, and to issue the declaratory relief and preliminary injunction, demonstrates what must occur if there is any hope of achieving social equity.

As previously noted, the victory for the environmental justice movement that was achieved in *South Camden* was short-lived (Garland 2007, 22–23). Five days after the decision in New Jersey, the U.S. Supreme Court issued its decision in *Alexander v. Sandoval* (2001) that there is no private right of action for individuals to sue the government under Section 602. The court said, however, that individuals may sue under Section 601, which prohibits intentional discrimination (*Alexander v. Sandoval* 2001, 282–286). Nevertheless, the court noted that a private right to enforce disparate impact regulations did not exist under Section 601. As a result, *South Camden* was not a victory for the environmental justice movement in the case of either Section 601 or 602.

It must be acknowledged that some U.S. circuit courts of appeal, including the Tenth Circuit, do not read *Alexander v. Sandoval* to preclude all claims brought to enforce Section 602 regulations (e.g., *Robinson v. Kansas* 2002). Rather, the Court's *Alexander* decision has been read to bar only disparate-impact claims brought by private parties directly under Title VI. The Tenth Circuit has stated that disparate-impact claims may still be brought against state officials under 42 U.S.C. § 1983 to enforce Section 602 regulations (*Robinson v. Kansas* 2002). To sustain a Section 1983 claim, the plaintiff must prove that the act complained of was committed by a person acting under color of state law and that such conduct deprived the plaintiff of a right, privilege, or immunity secured by the Constitution or the laws of the United States (see *Wilder v. Virginia Hospital Assoc.* 1990). Specifically, plaintiffs must prove that (1) that they are the intended beneficiary of the provision sought to be enforced; (2) that the right asserted is not so "vague and amorphous" that its enforcement strains judicial competence; and (3) that the provision unambiguously imposes a binding obligation on the state (see *Lucero v. Detroit Public Schools* 2001). Nevertheless, the circuits are split concerning whether *Alexander v. Sandoval* effectively precludes a Title VI plaintiff from bringing suit under Section 1983. Indeed, the Third, Sixth, and Ninth Circuits have interpreted that case to mean that plaintiffs cannot enforce regulations promulgated pursuant to Section 602 through a private cause of action under 42 U.S.C. § 1983 (see *Save Our Valley v. Sound Transit* 2003; *South Camden* 2001; *Wilson v. Collins* 2008).

In sum, Section 601 of Title VI requires proof of intentional discrimination and, although Section 602 does not necessarily require proof of intentional discrimination, the U.S. Supreme Court curtailed any private cause of action under Section 602 in *Alexander v. Sandoval*. As a result, Title VI is largely ineffective for environmental justice plaintiffs to seek justice in the courts. Still, in some circuits, the door is open for plaintiffs to pursue Title VI claims under Section 1983. To the extent that plaintiffs are able to prove discriminatory intent, Title VI *could* provide a useful framework for pursuing environmental justice claims, regardless of whether the court recognizes a plaintiff's right to pursue a private right of action under Section 602. As a result, one specific recommendation that environmental justice activists have been calling for is to amend Title VI so that it clearly allows for a private right of action in the case of any discriminatory outcome (Lawyers' Committee for Civil Rights under Law 2010). Such an approach would make the courts more effective at promoting environmental justice.

EPA Handling of Title VI Complaints

Outside the courts, there is also controversy regarding the EPA's failure to use Title VI to fight environmental injustice. A more extensive agency effort would occur in the agency itself through the EPA's Office of Civil Rights (OCR). The OCR does accept Title VI complaints, but it emphasizes that it does not represent "recipients" or "complainants" in the process (http://www.epa.gov/civilrights/docs/t6guidefaq2.pdf). Thus, while EPA Title VI complaints would not be decided in court, it is possible that they could eventually be adjudicated in courts upon appeal outside the agency. Moreover, some Title VI complaints may be linked to civil and criminal violations, so they could be adjudicated in federal court in the traditional enforcement process (Hiar 2011).

Unfortunately, Title VI complaints that have been filed with the OCR do not appear to be taken seriously by the EPA (Lawyers' Committee for Civil Rights under Law 2010). None of the hundreds of Title VI complaints, for instance, have been recognized as requiring that federal funding be withdrawn (Deloitte Consulting 2011; Hiar 2011). As the Lawyers' Committee for Civil Rights under Law (2010) points out, the total number of Title VI cases that are dismissed or rejected by the EPA is simply extraordinary.

A report produced for the EPA by Deloitte Consulting (2011, 15–16) suggests that this lack of attention to Title VI cases is a reflection of the fact that the OCR "lost sight of its mission and priorities." As one example of this assertion, Deloitte Consulting indicates that Title VI complaints are backlogged as long as four years. Specifically, the failure of the OCR to act on Title VI cases has put the agency at odds with the environmental justice ideals set out in EO 12898. To address this issue, the agency should seek to provide more resources to Title VI complaints to avoid delayed justice (Deloitte Consulting 2011). The OCR may also consider working more closely with private parties to provide guidance and resources that ensure that cases are successfully brought forward to help eliminate environmental injustice, as opposed to adopting a hands-off approach to the investigative process, which leads to the vague and inconsistent handling of complaints and does little to reduce environmental injustice.

Equal Protection Clause

Environmental justice plaintiffs have also focused on the Equal Protection Clause of the Fourteenth Amendment of the U.S. Constitution. The Equal Protection Clause provides, "No State shall … deny to any person within its jurisdiction the equal protection of the laws." In order to show a violation of the Equal Protection Clause, plaintiffs must demonstrate that defendants' actions resulted in a discriminatory impact and that the defendants intentionally or purposefully discriminated against them "based upon plaintiffs' membership in a protected class" (*Committee Concerning Community Improvement v. City of Modesto* 2009, 702–703).

The U.S. Supreme Court has recognized that "an invidious discriminatory purpose may often be inferred from the totality of the relevant facts, including the fact, if it is true, that the [policy] bears more heavily on one race than another" (*Washington v. Davis* 1976, 242). The Court further recognized that the "impact of an official action is often probative of why the action was taken in the first place since people usually intend the natural consequences of their actions" (*Reno v. Bossier Parish School Bd.* 1997, 487). But a facially neutral law is not necessarily invalid just because it "may affect a greater proportion of one race than of another. Disproportionate impact is not irrelevant, but it is not the sole touchstone of an invidious racial discrimination forbidden by the Constitution" (*Washington v. Davis* 1976, 242).

To show that facially neutral conduct is intentionally discriminatory in violation of the Equal Protection Clause, a plaintiff must demonstrate that the government "selected or reaffirmed a particular course of action at least in part 'because of,' not merely 'in spite of,' its adverse effects upon an identifiable group"(*Personnel Administrator of Massachusetts v. Feeney* 1979, 279). For example, the plaintiff may show that the policy was "applied in a discriminatory manner" or that it was adopted out of discriminatory animus (*South Camden* D.N.J. 2006, citing *Yick Wo v. Hopkins* 1886, 373–374; *Hunter v. Underwood* 1985). "Determining whether invidious discriminatory purpose was a motivating factor [in the adoption of a facially neutral policy] demands a sensitive inquiry into such circumstantial and direct evidence of intent as may be available" (*Village of Arlington Heights v. Metropolitan Housing Development Corp.* 1977, 266). In addition, the court may consider the historical background of the decision; the sequence of events leading up to the decision; any departures from the normal procedural or substantive sequence; the legislative or administrative history, including statements by decision-making officials; and the foreseeability of any disparate impact of the action (266; *see also Columbus Bd. Of Education v. Penick*, 1979, 464).

It is important to note that when a challenged governmental policy is "facially neutral," plaintiffs may also demonstrate discriminatory intent, in both Title VI and Equal Protection cases, by showing "gross statistical disparities" that show that "some invidious or discriminatory purpose underlies the policy" (*Committee Concerning Community Improvement v. City of Modesto* 2009, 703). Indeed, the courts have recognized that a showing of "gross statistical disparities" alone may provide sufficient proof of intent to discriminate (703). But such cases are rare; generally, plaintiffs must show more than just a discriminatory impact. Rather, as discussed previously, they should offer other evidence of discriminatory purpose or intent, such as historical information and legislative or administrative history (*Committee Concerning Community Improvement v. City of Modesto* 2009).

In an Equal Protection challenge, once plaintiffs meet the burden of establishing a discriminatory purpose based on race, that burden then shifts to the government to demonstrate that even in the absence of discriminatory animus, the same decision would have resulted (see *South Camden* D.N.J. 2006).

The Equal Protection Clause may, at first glance, appear to be a viable way for environmental justice plaintiffs to successfully challenge a discriminatory governmental action, especially given the potential under this clause to prove discriminatory intent through a showing of discriminatory impact. In practice, however, plaintiffs have had little success challenging governmental actions based on an Equal Protection argument because plaintiffs have difficulty establishing intent. For example, in *Bean v. Southwestern Waste Management Corp.* (1979), plaintiffs challenged the issuance of a landfill permit on the grounds that the governmental agency was behaving in a discriminatory fashion when granting permits (Garland 2007). The court found that although the three data sets that were offered "at first blush, look[ed] compelling, ... these statistics [broke] down under closer scrutiny" and ultimately, the plaintiffs were unable to prove discriminatory intent (*Bean v. Southwestern Waste Management Corp.* 1979, 678). *East Bibb Twiggs Neighborhood Association v. Macon Bibb Planning & Zoning Commission* (1989) is another example of the courts' refusing to find intentional discrimination when plaintiffs have asserted an Equal Protection claim after the siting of a landfill facility. (Garland 2007). The court found that while the landfill was located in a minority community, the decision was not motivated by a discriminatory intent. Because one landfill also existed in a Census tract with a majority white population, the court could not find a "clear pattern unexplainable on grounds other than race" under the *Village of Arlington Heights* standard (*East Bibb Twiggs Neighborhood Association v. Macon Bibb Planning & Zoning Commission* 1989). As Garland (2007, 17) recognizes, the *Bean* and *Twigg* cases "illustrate that proving discriminatory intent is a highly challenging standard of proof to meet." As a result of these strict and unrealistic standards of direct evidence of discrimination, there is not much hope that environmental injustice related to the distribution of hazards will be attenuated through court intervention and application of the Equal Protection Clause.

National Environmental Policy Act

Environmental justice plaintiffs have also attempted to assert claims under NEPA on the grounds that some federal agency acted in a biased, arbitrary, or capricious way. NEPA requires that federal agencies "take a 'hard look' at the environmental consequences" of a major federal action before taking such action (*Baltimore Gas & Elec. Co. v. Natural Resources Defense Council, Inc.* 1983). NEPA does not require a particular result and does not require that no environmental harm occur; rather, it imposes a mandatory review process (*Lujan v. Defenders of Wildlife* 1992, 605). Agencies must prepare a "detailed statement" so that a reviewing court may determine whether the agencies have made good-faith efforts to consider the NEPA values (National Environmental Policy Act of 1969, 42 U.S.C.§ 4332(1)(C)). Federal agencies must prepare an environmental impact statement (EIS) for "major Federal actions significantly affecting the quality of the human environment" [National Environmental Policy Act of 1969, 42 U.S.C.§ 4332(1)(C)], but they may first prepare an environmental assessment (EA) to determine whether an EIS is required [Natural Environmental Policy Act of 1969, 40 C.F.R. § 1508.9(a)(1)]. The EA is a concise document that provides enough evidence and analysis to allow determination of whether to prepare an EIS or a finding of no significant impact [Natural Environmental Policy Act of 1969 40 C.F.R. § 1508.9(a)(1)]. For example, environmental justice researchers and plaintiffs have alleged that an agency failed to properly consider environmental justice concerns in an EA prepared in connection with its NEPA review of a particular project (see Outka 2006; *One Thousand Friends of Iowa v. Mineta* 2002; *Saint Paul Branch of the NAACP v. U.S. Dept. of Transportation* 2011).

One of the bigger challenges for environmental justice plaintiffs in NEPA cases is establishing standing (*Saint Paul Branch of the NAACP v. U.S. Dept. of Transportation* 2011). That is, do environmental justice organizations lack standing when harms cannot be directly described and an agency determines that the project will have no significant impact on the environment? In order to establish standing, plaintiffs must show (1) that at least one of its members has suffered a concrete injury that is actual or imminent, not hypothetical; (2) that the injury is traceable to the defendant's action; and (3) that it is likely that the injury will be redressed by a favorable court decision (see *Friends of the Earth, Inc. v. Laidlaw Environmental Services* 2000). However, as a bar to standing, it has been held that "any obligation of [the federal agency] to consider environmental justice is not judicially enforceable" (*One Thousand Friends of Iowa v. Mineta* 2002, 1071). Further, it has been acknowledged by some federal district courts that environmental justice is not a right that can be used to appeal a federal agency decision (1084; see also *ACORN v. U.S. Army Corps of Engineers* 2000). Finally, it has been stated that "the failure to consider 'environmental justice' in and of itself cannot support a finding [that the federal agency] acted in an arbitrary and capricious manner" (*One Thousand Friends of Iowa v. Mineta* 2002, 25608 at *21; see also *Sur Contra La Contaminacion v. E.P.A.* 2000).

Other courts, however, have acknowledged that even though EO 12898 does not create a private right to judicial review, an action may still be subject to environmental justice review under NEPA and the Administrative Procedures Act. For example, the District of Columbia Court of Appeals addressed whether the Federal Aviation Administration (FAA) adequately considered environmental justice concerns in its review of an airport expansion plan (*Communities Against Runway Expansion, Inc. v. Federal Aviation Administration* 2004). In that case, the city of Boston initiated the environmental justice claim, arguing that the FAA's use of the entire county as the "potentially affected area," rather than the circumscribed greater Boston metropolitan area, was unreasonable. While the court disagreed, and thus the environmental justice claim failed, it bears noting that some courts will entertain environmental justice–based NEPA claims. Similarly, in *Mid States Coalition for Progress v. Surface Transportation Board* (2003), the Eighth Circuit Court of Appeals entertained an environmental justice–based NEPA claim and expressly recognized that "the purpose of an environmental justice analysis is to determine whether a project will have a disproportionately adverse effect on minority and low-income populations" (*Mid States Coalition for Progress v. Surface Transportation Board* 2003). That court further noted that "[t]o accomplish this, an agency must compare the demographics of an affected population with demographics of a more general character (for instance, those of an entire state)" (541). The Fifth Circuit Court of Appeals has also recognized that although EO 12898 does not create a private right of action, an agency's consideration of environmental justice issues is subject to review under the Administrative Procedure Act's "arbitrary and capricious" standard (*Coliseum Square Association, Inc. v. Jackson* 2006, 232).

In short, NEPA may, theoretically, be a viable basis on which to successfully assert an environmental justice claim. That is, if plaintiffs can establish that the federal agency was arbitrary or capricious in considering and disclosing the environmental justice impacts of a proposed action, they may be able to prevail in court. While this has not been tested in any court to date, there may be hope for court victories from this approach because it is increasingly clear that environmental justice is not examined in any systematic way across or within federal agencies. For example, Vajjhala, Van Epps, and Szambelan (2008) studied whether environmental justice issues increased in EISs and regulatory impact statements after EO 12898. While the researchers found that environmental justice issues were more likely to show up in EIS reports as a result of EO 12898, they also found high levels of inconsistency in the way that environmental justice was addressed in those statements. They also discovered that reports were arbitrary in the way they address environmental justice and did not "contain enough data to assess whether EJ impacts are significant" (Vajjhala, Van Epps, and Szambelan 2008, 2). In one agency (the Department of Transportation) the use of "environmental justice" actually decreases over time (Vajjhala, Van Epps, and Szambelan 2008). It should not be surprising, then, that environmental justice activists have recommended that NEPA be modified to require consistency and "identify environmental justice as an issue in NEPA compliance documents" (Lawyers' Committee for Civil Rights under Law 2010, 17). Federal agencies would then use EIS reports to ensure that environmental justice issues are monitored and that injustice is mitigated. Even if a plaintiff were to successfully assert an environmental justice claim under NEPA, such a win would not necessarily stop environmental injustice, but the agency would likely reconsider the impacts of such action on the plaintiff before making its decision.

Conclusions and Moving Forward

In this chapter, we reviewed environmental justice policy and litigation in the federal courts. We argue that courts are in a unique position to either increase or reduce environmental injustice through their sentencing duties and common law interpretation functions. In short, courts may help shape the environmental justice landscape. In the case of administrative, civil, and criminal adjudication for environmental violations, we find that courts have done little to ameliorate the problem of environmental injustice. However, it also appears that courts have done little to contribute to the problem. Overall, the studies by environmental justice academics appear to suggest that the monetary fines and penalties handed down from the courts are not discriminatory. Instead, these penalties may be the result of complex factors that include the types of violation that is pursued by regulatory agencies. If discrimination in sentencing exists, we suggest that it is indirect, and it may be more appropriately attributed to the prosecutor or state regulatory attorney who brings those charges to the courts. There is considerable precedence for this assertion in the criminology literature, where it has been well established that prosecutors have significantly more power than judges in determining the outcome of any particular case (Bibas 2003). We, therefore, propose that more attention be focused on prosecutorial discretion when examining environmental enforcement.

In the case of facility siting, we also find that federal courts have done little to combat environmental injustice. We recognize that Title VI has not been used as President Clinton proposed in EO 12898. Moreover, the EPA appears to continually reemphasize the importance of EO 12898 and the use of Title VI, but it does little to ensure environmental justice in the courts. Recently, the EPA (2011b) suggested that the agency is "pursuing long overdue, vigorous, robust, and effective implementation of Title VI." To date, this enforcement of Title VI in the courts has yet to take place in any serious fashion. Importantly, few Title VI cases have even been brought. Thus, there is really very little that courts do that promote equality under Title VI. As we have noted, the courts appear to rule in ways that facilitate and support indirect forms of discrimination because they refuse to acknowledge the disparate impact in permitting and siting decisions. As a result, we suggest that courts focus more on statistical evidence of indirect discrimination in overburdened communities, which should prove especially beneficial for those communities that already face disproportionate hazards. This might mean that the EPA identifies communities where environmental hazards exist and ensures that no additional hazards are sited in those communities. It may also mean that enforcement agencies should spend more time inspecting facilities that operate in communities that face significant hazards.

While the DOJ has supported a right of private enforcement of Title VI regulations, we find little evidence that it has continued to develop proactive policies that address environmental justice through the use of the courts in any significant way (Bratspies et al. 2008). Instead, the DOJ has focused on drug enforcement and gang activities in minority communities through anti-drug programs such as "Weed and Seed." This means that environmental hazards, including lead, will continue to threaten low-income communities. DOJ, in combination with the EPA, should redirect its efforts toward ensuring equality in environmental enforcement—especially in terms of prosecution. This approach might begin with a focus on lead poisoning and children, finding those areas where empirical evidence suggests that lead poisoning still presents a serious threat to low-income and minority children and may cause future problems such as learning disabilities, deviance, and even crime (Grandjean 2013). However, to achieve more environmental justice through the courts, any approach will need to go beyond simply talking about environmental justice in department documents. Instead, the DOJ might identify those communities that have the greatest need for enforcement efforts in the form of investigation and prosecution. We believe that such a proactive approach will bring more attention to environmental injustice in the courts and help facilitate corrective justice (see also Konisky 2009b). While additional efforts for achieving environmental justice are being pursued in *Plan EJ 2014*, these efforts have yet to produce changes in the courts, which go notably unmentioned in that planning document (EPA 2011).

References

Atlas, Mark. 2001. Rush to Judgment: An Empirical Analysis of Environmental Equity in U.S. Environmental Protection Agency Enforcement Actions. *Law & Society Review* 35 (3): 633–682.

Bibas, Stephanos. 2003. The Feeney Amendment and the Continuing Rise of Prosecutorial Power to Plea Bargain. *Journal of Criminal Law & Criminology* 94 (2): 295–308.

Bonner, Kathleen. 2012. Toxins Targeted at Minorities: The Racist Undertones of Environmentally-Friendly Initiatives. *Villanova Environmental Law Journal* 23:89–115.

Bowers, Meredith J. 1995. The Executive's Response to Environmental Injustice: Executive Order 12898. *Environmental Law (Northwestern School of Law)* 1 (2): 645–665.

Bratspies, Rebecca M., David M. Driesen, Robert L. Fischman, Sheila Foster, Eileen Gauna, Robert L. Glicksman, Alexandra B. Klass. 2008. Protecting Public Health and the Environment by the Stroke of a Presidential Pen: Seven Executive Orders for the President's First 100 Days. White Paper #806. Center for Progressive Reform. Available at http://www.progressivereform.org/CPR_Exec Orders_Stroke_of_a_Pen.pdf.

Browner, Carol M. 1995. *The Environmental Protection Agency's Environmental Justice Strategy.* http://www.epa.gov/environmentaljustice/resources/policy/ej_strategy_1995.pdf.

Bryant, Bunyan. 1995. Issues and Potential Policies and Solutions for Environmental Justice: An Overview. In *Environmental Justice: Issues, Policies, and Solutions*, ed. Bunyan Bryant, 1–7. Washington, DC: Island Press.

Bullard, Robert D. 1983. Solid Waste Sites and the Black Houston Community. *Sociological Inquiry* 53: 273–288.

Bullard, Robert D. 1990. *Dumping in Dixie: Race, Class, and Environmental Quality.* Boulder, CO: Westview.

Bullard, Robert D., and Glenn S. Johnson. 2000. Environmental Justice: Grassroots Activism and Its Impact on Public Policy Decision Making. *Journal of Social Issues* 56 (3): 555–578.

Burns, Ronald G., Michael J. Lynch, and Paul Stretesky. 2009. *Environmental Law, Crime, and Justice.* El Paso, TX: LFB Scholarly.

Clinton, William. 1994. Memorandum for the Heads of All Departments and Agencies. http://www.epa.gov/compliance/ej/resources/policy/clinton_memo_12898.pdf.

Cole, Luke W. 1991. Remedies for Environmental Racism: A View from the Field. *Michigan Law Review* 90:1991–1997.

Cole, Luke W. 1993. Environmental Justice Litigation: Another Stone in David's Sling. *Fordham Urban Law Journal* 21 (3): 523–545.

Cole, Luke W., and Sheila R. Foster. 2001. *From the Ground Up: Environmental Racism and the Rise of the Environmental Justice Movement.* New York: New York University Press.

Davis, Angela J. 1998. Prosecution and Race: The Power and Privilege of Discretion. *Fordham Law Review* 67: 13–68.

Deloitte Consulting. 2011. *Evaluation of the EPA Office of Civil Rights.* Deloitte Consulting. Washington, DC. http://www.epa.gov/epahome/pdf/epa-ocr _20110321_finalreport.pdf.

Environmental Protection Agency (EPA). 1998. *Interim Guidance for Investigating Title VI Administrative Complaints Challenging Permits.* February. Washington, DC. http://www.enviro-lawyer.com/Interim_Guidance.pdf.

Environmental Protection Agency (EPA). 2011. *Response to Public Comments on Plan EJ 2014 Strategy and Implementation Plans.* Washington, DC. http://www .epa.gov/environmentaljustice/plan-ej/.

Environmental Protection Agency (EPA). 2013. Plan EJ 2014 Progress Report. Washington, DC. http://www.epa.gov/environmentaljustice/resources/policy/plan -ej-2014/plan-ej-progress-report-2013.pdf.

Garland, Meagan Elizabeth Tolentino. 2007. Addressing Environmental Justice in Criminal Sentencing Process: Are Environmental Justice Communities "Vulnerable Victims'" under 3A1.1(B)(1) of the Federal Sentencing Guideline in the Post–*United States v. Booker* Era? *Albany Law Environmental Outlook Journal* 12 (1): 1–51.

General Accounting Office (GAO). 1983. *Siting of Hazardous Waste Landfills and Their Correlation with Racial and Economic Status of Surrounding Communities. GAO/RCED-83-168.* Washington, DC: Government Printing Office.

Godsil, Rachel. 1991. Remedying Environmental Racism. *Michigan Law Review* 90:394–427.

Gould, Kenneth A., David N. Pellow, and Allan Schnaiberg. 2008. *The Treadmill of Production: Injustice and Unsustainability in the Global Economy.* Boulder, CO: Paradigm.

Greife, Matthew. 2012. "Environmental Crime: An Empirical Investigation into Factors Influencing Legal Outcomes." Master's thesis, University of Colorado, Denver.

Grandjean, Philippe. 2013. *Only One Chance: How Environmental Pollution Impairs Brain Development—and How to Protect the Brains of the Next Generation.* New York: Oxford University Press.

Hammer, Natalie M. 1996. Title VI as a Means of Achieving Environmental Justice. *Northern Illinois University Law Review* 16: 693–715.

Hiar, Corbin. 2011. Environmental Injustice: EPA Neglects Discrimination Claims from Polluted Communities. iWatch News: The Center for Public Integrity. http:// www.publicintegrity.org/2011/12/14/7660/environmental-injustice-epa-neglects -discrimination-claims-polluted-communities.

Konisky, David M. 2009a. The Limited Effects of Federal Environmental Justice Policy on State Enforcement. *Policy Studies Journal* 37 (3): 475–496.

Konisky, David M. 2009b. Inequities in Enforcement? Environmental Justice and Government Behavior. *Journal of Policy Analysis and Management* 28 (1): 102–121.

Lavelle, Marianne, and Marcia Coyle. 1992. Unequal Protection: The Racial Divide in Environmental Law. *National Law Journal* 15 (3): S1–S12.

Lawyers' Committee for Civil Rights under Law. 2010. Environmental Injustice in the U.S. and Recommendations for Eliminating Disparities. Washington, DC. Available at http://www.lawyerscommittee.org/admin/site/documents/files/Final-Environmental-Justice-Report-6-9-10.pdf.

Lynch, Michael J., Paul B. Stretesky, and Ronald G. Burns. 2004. Determinants of Environmental Law Violation Fines against Petroleum Refineries: Race, Ethnicity, Income, and Aggregation Effects. *Society & Natural Resources* 17 (4): 333–347.

Mank, Bradford C. 1999a. Environmental Justice and Title VI: Making Recipient Agencies Justify Their Siting Decisions. *Tulane Law Review* 73:787–843.

Mank, Bradford C. 1999b. Is There a Private Cause of Action Under EPA's Title VI Regulations?: The Need to Empower Environmental Justice Plaintiffs. *Columbia Journal of Environmental Law* 24:1–61.

Mank, Bradford C. 2008. Title VI. In *The Law of Environmental Justice: Theories and Procedures to Address Disproportionate Risks*, ed. Sheila R. Foster, 23–65. Chicago: American Bar Association.

McSpadden, Lettie. 1995. The Courts and Environmental Policy. In *Environmental Politics and Policy: Theories and Evidence*, ed. James Lester, 242–274. Durham, NC: Duke University Press.

Melnick, R. Shep. 1983. *Regulation and the Courts: The Case of the Clean Air Act*. Washington, DC: Brookings Institution Press.

Mennis, Jeremy L. 2005. The Distribution and Enforcement of Air Polluting Facilities in New Jersey. *Professional Geographer* 57 (3): 411–422.

Meyer, David S. 2004. Protest and Political Opportunities. *Annual Review of Sociology* 30:125–145.

Murphy-Greene, Celeste, and Leslie A. Leip. 2002. Assessing the Effectiveness of Executive Order 12898: Environmental Justice for All? *Public Administration Review* 62 (6): 679–687.

New York State Department of Environmental Conservation. 2003. Commissioner Policy 29, Environmental Justice and Permitting. Available at http://www.dec.ny.gov/regulations/36951.html.

O'Leary, Rosemary. 1993. *Environmental Change: Federal Courts and the EPA*. Philadelphia: Temple University Press.

O'Neil, Sandra G. 2007. Superfund: Evaluating the Impact of Executive Order 12898. *Environmental Health Perspectives* 115 (7): 1087–1093.

Outka, Uma. 2006. NEPA and Environmental Justice: Integration, Implementation, and Judicial Review. *Boston College Environmental Affairs Law Review* 33 (3): 601–626.

Ringquist, Evan J. 1998. A Question of Justice: Equity in Environmental Litigation, 1974–1991. *Journal of Politics* 60:1148–1165.

Ringquist, Evan J. 2004. Environmental Justice. In *Environmental Governance Reconsidered*, ed. Robert F. Durant, Daniel J. Fiorino, and Rosemary O'Leary, 225–288. Cambridge, MA: MIT Press.

Simms, Patrice Lumumba. 2013. On Diversity and Public Policymaking: An Environmental Justice Perspective. *Sustainable Development Law & Policy* 13 (1): 14–19.

Stretesky, Paul, and Michael J. Hogan. 1998. Environmental Justice: An Analysis of Superfund Sites in Florida. *Social Problems* 45 (2): 268–287.

Taylor, Dorceta E. 2000. The Rise of the Environmental Justice Paradigm: Injustice Framing and the Social Construction of Environmental Discourses. *American Behavioral Scientist* 43 (4): 508–580.

U.S. Commission on Civil Rights. 2003. *Not in My Backyard: Executive Order 12898 and Title VI as Tools for Achieving Environmental Justice. October.* Washington, DC: U.S. Commission on Civil Rights.

Vajjhala, Shalini, Amanda Van Epps, and Sarah Szambelan. 2008. *Integrating EJ into Federal Policies and Programs.* Washington, DC: Resources for the Future.

Weinrib, Ernest J. 2012. *The Idea of Private Law.* Oxford, UK: Oxford University Press.

U.S. Supreme Court Cases

Alexander v. Sandoval, 532 U.S. 275 (Westlaw 2001)

Wilder v. Virginia Hospital Assoc., 496 U.S. 498 (Westlaw 1990)

Washington v. Davis, 426 U.S. 229 (Westlaw 1976)

Reno v. Bossier Parish School Bd., 520 U.S. 471 (Westlaw 1997)

Personnel Administrator of Massachusetts v. Feeney, 442 U.S. 256 (Westlaw 1979)

Yick Wo v. Hopkins, 118 U.S. 356 (Westlaw 1886)

Hunter v. Underwood, 471 U.S. 222 (Westlaw 1985)

Village of Arlington Heights v. Metropolitan Housing Development Corp. (Westlaw 1977)

Columbus Board of Education v. Penick, 443 U.S. 449 (Westlaw 1979)

Lujan v. Defenders of Wildlife, 504 U.S. 555 (Westlaw 1992)

Baltimore Gas & Elec. Co. v. Natural Resources Defense Council, Inc., 462 U.S. 87 (Westlaw 1983)

Friends of the Earth, Inc., v. Laidlaw Envtl. Servs., 528 U.S. 167 (Westlaw 2000)

U.S. Court of Appeals Cases

Save Our Valley v. Sound Transit, 335 F.3d 932 (9th Cir. Westlaw 2003)

Committee Concerning Community Improvement v. City of Modesto, 583 F.3d 690 (9th Cir. Westlaw 2009)

South Camden Citizens in Action v. New Jersey Dep't of Env. Protection, 274 F.3d 771 (3d Cir. Westlaw 2001)

Wilson v. Collins, 517 F.3d 421 (6th Cir. Westlaw 2008)

Robinson v. Kansas, 295 F.3d 1183 (10th Cir. Westlaw 2002)

Sur Contra La Contaminacion v. E.P.A., 202 F.3d 443 (1st Cir. Westlaw 2000)

Mid States Coalition for Progress v. Surface Transportation Board, 345 F.3d 520 (8th Cir. Westlaw 2003)

Coliseum Square Association, Inc. v. Jackson, 465 F.3d 215 (5th Cir. Westlaw 2006)

U.S. District Court Cases

South Camden Citizens in Action v. NJDEP, 145 F.Supp.2d 446 (D.N.J. Westlaw 2001)

South Camden Citizens in Action v. NJDEP, 2006 U.S. Dist. LEXIS 45765, *70 (D.N.J. LexisNexis 2006)

Lucero v. Detroit Public Schools, 160 F.Supp.2d 767 (E.D. Mich. Westlaw 2001)

Bean v. Southwestern Waste Management Corp., 482 F.Supp. 673 (S.D. Tex. Westlaw 1979)

East Bibb Twiggs Neighborhood Ass'n v. Macon Bibb Planning & Zoning Commission, 706 F. Supp. 880 (M.D. Ga. Westlaw 1989)

One Thousand Friends of Iowa v. Mineta, 250 F. Supp. 2d 1064 (D. Iowa Westlaw 2002)

One Thousand Friends of Iowa v. Mineta, 250 F. Supp. 2d 1075, 1084 (D. Iowa 2002)

One Thousand Friends of Iowa v. Mineta, 2002 U.S. Dist. LEXIS 25608 (D. Iowa 2002)

Saint Paul Branch of the NAACP v. U.S. Dept. of Transportation, 764 F.Supp. 2d 1092 (D. Minn. Westlaw 2011)

ACORN v. U.S. Army Corps of Engineers, 2000 U.S. Dist. LEXIS 5408 (E.D.La. LexisNexis 2000)

Communities Against Runway Expansion, Inc., v. Federal Aviation Administration, 355 F.3d 678 (D.C. Cir. Westlaw 2004)

9

Federal Environmental Justice Policy: Lessons Learned

David M. Konisky

This book has taken a critical look at federal environmental justice policy as it has been implemented over the past twenty years. The time is long overdue for such an analysis. Two decades have passed since President Bill Clinton signed Executive Order 12898 (EO 12898), the hallmark federal action to address environmental justice. Yet the scholarly literature has not thoroughly examined how the federal government, particularly the Environmental Protection Agency (EPA), has performed in translating the principles and aspirations of EO 12898 into action. Given the breadth of regulatory activity of the EPA, it is important to evaluate the efficacy of the agency's response, and specifically whether it has successfully integrated consideration of environmental equity into its day-to-day decision making.

Concerns about environmental inequity rose to prominence thirty years ago, mostly in light of siting controversies over hazardous waste landfills and incinerators in poor and minority communities, urban and rural alike. But such controversies recur today in the context of many different types of locally unwanted land uses. One of the central fronts of this conflict has become power plants. For example, a grassroots movement backed by several elected officials in Brockton, Massachusetts (a city of about 95,000 located about 25 miles south of Boston) has been fighting the construction of a new gas-fired electric power plant for much of the last decade. The $350 million facility was approved by the state Energy Facilities Siting Board in September 2011, but it has continued to face stiff opposition from members of the community. Environmental justice advocates have argued that residents of Brockton, which has a minority population of nearly 40 percent and has a median household income about 25 percent less than the national average, already experience significant environmental risks from existing hazardous waste sites and major sources of toxic and conventional air pollution (Faber 2008). Environmental justice and civil rights organizations are protesting similar power plant projects across the country, as well as advocating for the closure of existing plants (Wilson 2012). Other issues also increasingly garner the attention of the environmental justice movement (in addition, to their traditional emphasis on hazardous waste facilities), ranging from local problems, such as the effects of concentrated animal feeding operations and the availability of urban green space, to global ones, such as the effects of climate change.

Although the causes of environmental inequities remain hotly disputed, there remains strong evidence of race- and income-based disparities in environmental burdens, ranging from the location of commercial hazardous waste treatment, storage, and disposal facilities (Bullard et al. 2007) to poor air quality (Bell and Ebisu 2012; Clark, Millet, and Marshall 2014; Miranda et al. 2011). Environmental justice challenges, therefore, are not just a thing of the past. They remain a pressing concern for the residents of many low-income and minority communities who rightfully worry that their health and general quality of life may be compromised by the environmental risks around them, and that government is not on their side when it comes to protecting them from these risks. Although undoubtedly much is still to be discovered regarding the nature and extent of environmental inequities and their economic, political, and social origins, this book has sought to turn the spotlight more brightly on the effectiveness (or lack thereof) of the policy response. Understanding the opportunities and challenges that government agencies face in addressing inequities is important for advancing the environmental justice research agenda beyond just documenting problems to devising potential solutions to those problems.

This concluding chapter has three specific objectives. First, I will summarize the central conclusions reached in the preceding chapters of this book. Second, I will distill a set of common factors that emerge from these analyses as impediments to the effective implementation of federal environmental justice policy. Third, I will draw some general conclusions about the opportunities and constraints facing the EPA as it advances its environmental justice policy agenda.

Some Key Conclusions

What do the analyses in this book reveal about the performance of the federal government, and specifically the EPA, in achieving its stated environmental justice goals? Has the agency successfully incorporated consideration of environmental justice in its decision making in the areas of permitting, standard-setting and rulemaking analysis, and public participation and enforcement? The authors have reached many conclusions on these different dimensions of environmental decision making.

Eileen Gauna concludes in chapter 3 that attention to equity considerations in permitting under major pollution control statutes has been limited and inconsistent. Although EO 12898 does not explicitly mention permitting, it does mandate that each federal agency identify and address "disproportionately high and adverse human health or environmental effects of its programs, policies, and activities on minority populations and low-income populations" (EO 12898), which certainly extends to permitting decisions. One of the key questions raised in Gauna's analysis is whether the EPA fully considers disproportionate impacts on vulnerable populations when it issues permits under laws such as the Clean Air Act (CAA) and the Resource Conservation and Recovery Act (RCRA). This is a challenging question to answer, given the sheer number of environmental permits issued by the EPA under these laws (not to mention the state administrative agencies that issue permits under EPA oversight).[1]

Gauna approaches this question first through an analysis of important permit cases adjudicated by the EPA's Environmental Appeals Board (EAB). One important finding from her analysis is that, to date, the EAB has not overturned an EPA-issued permit on the grounds that it results in disproportionate impacts for either low-income or minority populations. More precisely, there has yet to be a successful challenge by an environmental justice advocacy organization to a permit in cases ruled on by the EAB.[2] In this regard, the EAB has adopted a limited interpretation of the reach of EO 12898 and has decided that it will not make independent assessments about environmental inequities. Rather, the EAB has elected to defer to the judgment of EPA permit writers and their analysis of the distribution of impacts from a new permit. And, to date, the EPA when challenged has not concluded that its permits would result in inequitable outcomes for environmental justice communities.

The EAB, however, has set a procedural baseline for the EPA to follow when making its permit decisions. Since the permit renewal case for a Chemical Waste Management facility in Fort Wayne, Indiana, in 1995, the EAB has required that the agency include meaningful public participation and conduct appropriate environmental justice analysis. On its face, this is significant. Environmental justice advocates have long demanded that environmental permitting decisions be made with their input, and with the full consideration of possible impacts (including cumulative impacts) on vulnerable communities. In practice, however, the ways in which EPA permit writers have conducted environmental justice analysis has varied considerably, differing on almost a case-by-case basis. To date, the EAB has granted wide deference to the agency in this respect, requiring simply that such analysis be done, not that it be done in a specific manner. In the absence of clear guidance from EPA headquarters on how such analysis should be carried out, EPA regions have developed their own practices. As a consequence, there are no consistent parameters employed for identifying disparate impacts of new permits on low-income and minority populations, or even what type of communities should be considered when making such a determination.

An additional implication from Gauna's analysis is that, while the EPA has generally not pursued a strategy of writing permits in a way to explicitly achieve environmental justice goals, the EAB might grant it the latitude to do so. To date, the EAB has not ruled in favor of environmental justice challenges, but this is based on its deference to EPA analysis that the permits that it has issued do not create disparate impacts for low-income and minority populations. This does not necessarily preclude EPA permit writers from adjusting, if not denying outright, a permit on the grounds that it would result in such a disparate impact. The EPA has not pushed the envelope in this regard, despite having some discretion to do so under several environmental statutes (EPA Office of General Counsel 2000, National Academy of Public Administration 2001).[3]

Doug Noonan's analysis of EPA standard setting in chapter 4 finds sparse evidence that the EPA has systematically promulgated rules that seek to equalize pollution burdens across subgroups in the population. He concludes that, to this point, the agency "has done relatively little to spare particular subgroups from the burdens of regulations or of pollution." In this sense, EO 12898 and other agency policy commitments to achieve environmental justice do not seem to have resulted in a significant change to how EPA designs standards or sets emissions limits under major pollution control laws, at least until very recently. As Noonan cites in this chapter, Gina McCarthy, the current EPA administrator and then–assistant administrator for the Office of Air and Radiation (OAR), noted that the EPA's updated National Ambient Air Quality Standard (NAAQS) for NO_2, finalized in 2010, was "the very first time a rule actually recognized the issues of disproportional impact" (NEJAC 2010, p. 92). Although McCarthy touted the new NO_2 standards as a significant achievement for environmental justice, it should be underscored that the rule called for additional emissions monitoring in susceptible and vulnerable communities, not a differential limit on ambient concentrations for these communities. More generally, it also reflects a clear recognition that, in the twenty years since EO 12898, environmental justice is still not a major part of how standards are set by EPA regulators.

That said, it must be recognized that the agency's mandate under major federal pollution control statutes is not to protect specific populations, but rather to protect human health and the environment for everyone. In other words, the EPA is constrained in its authority by the statutes it implements. But, as Noonan argues, there are ways in which the EPA could use the discretion that it does have when choosing policy instruments and in deciding pollution standards. Although the agency must adhere to normal rulemaking procedures, Noonan concludes that "[p]romulgating regulations to favor environmental justice goals can still be accomplished in the right political circumstances, especially with creative rulemaking."

Even if environmental justice considerations have not led to changes in the design of new pollution control or emissions standards, to what extent have distributional effects been at least seriously considered in economic analyses of proposed new standards? This is the question that Ron Shadbegian and Ann Wolverton address in chapter 5. Economic analysis of major rules has been a critical (and contentious) component of EPA rulemaking since President Ronald Reagan issued Executive Order 12291, which required that such an analysis be done for any new major rule. Yet, as a matter of historical practice, Shadbegian and Wolverton note that few economic reviews have included a comprehensive assessment of distributional impacts, particularly in the way called for in EO 12898, where equity effects would be directly considered with respect to specific population groups (i.e., minority, low-income, and indigenous groups).[4]

This has begun to change in recent years, however, as there has been a significant uptick in the number of environmental justice evaluations conducted as part of economic analyses. But, as Shadbegian and Wolverton note, "With this increased attention on evaluating potential EJ concerns associated with rulemaking comes the necessity of grappling with a number of analytic issues raised in the academic literature." In their analysis of five recent rulemakings, they illustrate the different ways that the EPA has addressed complex questions regarding determination of the appropriate geographic scope, identification of potentially affected populations, selection of the right comparison groups, spatial identification of the effects on population groups, and development of proxies for exposure. Although these are not easy questions to answer, particularly in light of data and methodological limitations, the consideration of environmental justice issues as part of EPA economic analyses has been further hampered by a lack of technical guidance. As a consequence, when environmental justice is considered, there is considerable inconsistency in how it is done across rules, even within the same pollution control program.

In chapter 6, Dorothy Daley and Tony Reames evaluate how the EPA and two other federal agencies (the Department of Energy and the Department of Transportation) have done in improving public participatory processes in pursuit of environmental justice goals; their analysis reveals further inconsistency. For the EPA, clear successes lie with the establishment of formal institutions such as the National Environmental Justice Advisory Council (NEJAC). The NEJAC is a multi-stakeholder advisory group that has been a constant and constructive voice speaking in the ear of the EPA, bringing attention to various dimensions of environmental justice in agency decision making. In addition, Daley and Reames argue that the EPA's support of the Environmental Justice Small Grants Program has been an innovative way to help empower communities to address the challenges they confront, although they also note that this program requires a sustained commitment to be successful over the long run, something it has not always had.

Overall, Daley and Reames conclude that, although EO 12898 has led to more opportunities for public involvement in environmental decision making, "participation from low-income and minority communities remains uneven." They argue that the failure of the EPA to clearly delineate which groups need to be better represented in federal environmental decision making has impeded the development of effective participatory mechanisms. In addition, the fragmented nature of decision making at the EPA further complicates efforts to meaningfully involve stakeholders, especially for low-income and minority groups who tend to be less active in government decision making in general. As they note in their discussion, creating more mechanisms for the environmental justice community to participate in federal environmental decision making has been a high priority during the Barack Obama administration. Former administrator Lisa Jackson repeatedly referenced the need to "expand the conversation" on environmental policy. Whether the renewed focus is sustained remains to be seen.

The last two chapters of the book examine the enforcement stages of the regulatory process, first in terms of actions taken by the EPA and state administrative agencies to secure compliance with federal pollution control laws, and then in terms of the use of the courts by advocacy organizations and the EPA and the Department of Justice (DOJ) to adjudicate environmental justice challenges. In chapter 7, which I co-authored with Chris Reenock, we found little evidence that EPA enforcement practices changed in the years after EO 12898. Our analysis of CAA enforcement determined that regulated firms in poor and minority communities were no more likely to receive enforcement attention (such as compliance inspections and punitive actions for violations) in the years after the policy reforms were put in place than they were before. Similarly, we did not find much evidence that states responded to the federal policy signal by intensifying their enforcement efforts in these communities. These findings stand in contrast to the endorsement expressed in EO 12898 and EPA's commitment in its 1995 environmental justice strategy to target enforcement attention to facilities located in low-income and minority communities.[5]

Regulatory enforcement is an area of decision making where agencies have clear discretion to pursue actions to achieve environmental justice goals. There is no legal constraint against the government agencies aggressively pursuing compliance of firms located in environmental justice communities, and there is ample social science evidence to suggest that strong enforcement deters noncompliance (Cohen 1999; Gray and Shimshack 2011). In this sense, the EPA and the state administrative agencies that it oversees have not used the impetus of EO 12898 to shift their enforcement activities to pursue environmental justice goals.[6]

In chapter 8, Elizabeth Gross and Paul Stretesky provided a detailed assessment of a number of important issues concerning the handling of environmental justice issues in the courts. They reach several conclusions in their analysis. First, Gross and Stretesky evaluated the empirical literature that tests the claim that federal courts reach biased decisions by penalizing firms found to be violating pollution control laws less harshly when they are located in low-income and minority communities. This claim has become a virtual article of faith among many in the environmental justice advocacy community,[7] but their review of the most recent and carefully executed studies on the subject indicate only weak empirical evidence supporting this claim.[8]

They next turned to the use of the courts by environmental justice advocates, painting a rather bleak picture of the potential of successful challenges. Under Title VI of the federal Civil Rights Act, the Equal Protection Clause of the Fourteenth Amendment to the U.S. Constitution, and the National Environmental Policy Act (NEPA), challenges to federal environmental decision making have largely proved unsuccessful. Most of these cases have been brought in the context of facility siting and permitting, and Gross and Stretesky conclude that federal courts have repeatedly held that there is no disparate impact, in part because of the high legal burden that plaintiffs must overcome. Moreover, they conclude that the EPA and the DOJ have not aggressively used the courts to achieve environmental justice goals, despite the endorsement of such an approach by President Clinton, the EPA and the DOJ themselves, the NEJAC, and others. Finally, they note that the EPA itself has been very slow to resolve Title VI complaints brought to it by environmental justice advocates, a persistent problem for the agency for which it has received considerable attention and criticism.

Some Broad Themes

Overall, what can we take away from these analyses? Certainly, if one were to weigh the collective evidence presented in this book, there are many more shortcomings of policy than clear successes. The EPA in particular has not given environmental justice sufficient priority to convert the reforms of the mid-1990s into sustained changes in its decision making. In addition, where progress has been made, it has been slow and inconsistent across regulatory activities and programs. In the end, it is fair to conclude that the performance of the federal government on achieving environmental justice has been disappointing.

The modest impact of EO 12898 on federal environmental decision making illustrates the limits to pursuing policy solely through administrative actions such as executive orders. Recent scholarship in political science has highlighted a growing tendency of presidents to use administrative measures to achieve environmental policy goals, with several examples of success (Klyza and Sousa 2008). The case of federal environmental justice policy shows that the effectiveness of such measures is by no means certain.

To be sure, the task given to the EPA was a challenging one. The policy mandates expressed in EO 12898 were broad in scope and expectation, and they did not empower the EPA with additional authority or resources to ensure that the objectives would be fulfilled. Of course, the agency already had a full plate of responsibilities to address, and the failure to more fully respond to environmental inequities does not mean that the EPA has failed to effectively carry out its mission in other areas. In fact, one of the complicating factors for achieving environmental justice goals is that the statutes that give the EPA its marching orders do not demand consideration of equity. As a consequence, the environmental justice programs have to be built on top, and within the constraints, of the existing statutory infrastructure, and in a way so as not to compromise the EPA's more general mission to protect human health and the environment.

Although these are real obstacles to effective implementation of federal environmental justice policy, this book has also suggested ways that the EPA has come up short, even in areas where it has the capacity to act. Three factors in particular repeatedly surfaced throughout the chapters as major impediments to progress: (1) the failure to develop clear policy guidance, (2) poor coordination across EPA regions and states, and (3) inconsistent agency leadership. Each of these items is discussed in turn, although they are obviously interrelated.

The EPA has been perpetually slow (if not negligent) to develop policy and technical guidance. These types of guidance documents are critical for translating statutory language, regulatory programs, and general policy declarations into actionable procedures. In the case of environmental justice policy, they are especially important for converting the broad policy goals and aspirations of EO 12898 and other strategy documents into steps that permit writers, rulemakers, enforcement officials, and others should take to incorporate equity considerations into their programs and activities.[9] The lack of clear guidance has been an impediment in many of the areas of decision making analyzed in this book. Even in cases where the EPA did develop guidance, it took the agency many years to do it, and the guidelines developed often did not resolve key questions, resulting in significant delays to pursuing environmental justice goals.[10]

For example, take the EPA's guidance on how to perform an environmental justice assessment, which was not published until ten years after EO 12898. These assessments are necessary to determine whether a regulatory action—such as the siting or permitting of a new pollution source or the setting of a new pollution control standard—will generate disproportionate impacts for particular communities or segments of the population. According to the agency, the aim of its *Toolkit for Assessing Potential Allegations of Environmental Injustice (EJ Toolkit)*, was to provide a conceptual framework for understanding the EPA's environmental justice program as a whole, and to provide a systematic approach that could be used and adapted to assess and respond to current and future allegations of environmental injustice (EPA 2004).

Although in many respects, the *EJ Toolkit* was a comprehensive statement of how the EPA would consider environmental inequities, it was nonbinding on its future decision making. Moreover, the guidance sent mixed signals. On one hand, the document instructed: "On a day-to-day basis, program staff have the responsibility to seek ways to integrate environmental justice considerations into EPA's programs, policies, and activities. Therefore, program staff should use the framework presented in this document to promote national consistency in how environmental justice concepts are understood and addressed throughout the Agency" (EPA 2004, p. 4). On the other hand, the *EJ Toolkit* also indicated that "the decision on whether and how to use the tools and the methodology presented in this document should be made on a case-by-case basis" (p. 15), thereby offering decision makers discretion in not just how to use the recommended framework, but also whether such an analysis was necessary at all (in essence, undermining the very point of EO 12898). It should come as no surprise then that agency decisions have not regularly been accompanied by these types of assessments, as was documented in several chapters.

The *EJ Toolkit* also either muddled key concepts or left them undefined. For instance, the guidance made a point to emphasize that environmental justice assessments should not be limited to analyzing potential inequities confronting low-income and minority groups, but should be applied in any situation in which one geographic area or population subgroup experienced a potential disparate impact. Although one could certainly make an argument that any government decision or policy should not disproportionately affect one group or another, EO 12898 explicitly called for the protection of low-income and minority communities. More generally, the guidance left critical analytic decisions unaddressed. This point was emphasized by Shadbegian and Wolverton in the context of economic analysis, but the implications were also evident in the chapters on permitting, standard setting, and enforcement. Shadbegian and Wolverton highlighted key areas that require clarification: determining the right geographic scope to consider, identifying the potentially affected population and reasonable comparison groups, and determining the effects on population groups and proxying exposure. These are difficult concepts to define, but without clear guidance, that task is left to the discretion of individual decision makers (or the EPA regional or state offices in which they work). As a consequence, there has been immense variation in practice both within and across the programs managed by the EPA.

The *EJ Toolkit* was not designed to prescribe how decisions or actions could or should be modified in response to an environmental justice assessment. Unfortunately, for many areas of decision making, the EPA has not developed or put into use this type of formal guidance. In the past few years, the EPA has begun to rectify this situation. For example, the agency released its *Interim Guidance on Incorporating Environmental Justice During the Development of an Action* (*Interim Guidance*) in July 2010, which provided a framework for considering distributional issues throughout each stage of rulemaking (EPA 2010), and additional technical guidance is under development as part of *Plan EJ 2014*, discussed more later in this chapter. But, the very fact that the EPA is developing guidance to inform and direct decision making for its core regulatory functions for the first time speaks volumes about the agency's lack of attention to achieving the equity goals clearly articulated in EO 12898 and routinely espoused by agency leadership.

A second common factor that has impeded effective implementation of federal environmental justice policy has been the poor coordination of activities across federal and state agencies. This lack of coordination, of course, is directly related to the EPA's failure to develop clear national policy guidance and implementation procedures, but it also is more general, extending to the agency's oversight and management of its regional offices and state administrative agencies. Environmental protection in the United States is a shared responsibility across levels of government, and significant authority for policy implementation rests outside of Washington, D.C. As a matter of administrative necessity (as well as political factors at times), the EPA has delegated considerable authority to its ten regional offices and to state administrative agencies to implement many critical regulatory functions, including most permitting and enforcement under major federal pollution control laws.[11] With this delegation, however, comes the management challenge of ensuring that federal policy is carried out in a consistent manner across the country.

The lack of policy coordination was evident in several of the chapters. For example, permit writers in EPA regional offices performed environmental justice assessments using different assumptions about the affected population around a potentially permitted facility (e.g., populations 1 mile away, 2 miles away, and so on). Although these decisions in part reflected the specifics of each case, they were not grounded in a formal, transparent framework, leaving an impression that they were determined on an ad-hoc basis. The inconsistent way analyses were performed to support permit decisions points to a more general problem highlighted by the EPA Inspector General in its 2004 oversight report of the EPA's implementation of EO 12898. The Inspector General found that the EPA regional offices had, in the absence of guidelines and procedures from EPA headquarters, developed their own rules and policies on fundamental questions, such as defining the demographic criteria to qualify a community as an environmental justice area. In practice, the regions use diverse measures and thresholds of income and minority attributes, leading the Inspector General to conclude that "[t]hese disparate definitions have created inconsistencies among the regions as to who should be included in a defined environmental justice area" (p. 8).

Another important aspect of EPA policy coordination regards oversight and management of programs delegated to the states. The analysis of state enforcement of the federal CAA in chapter 7 does not provide much indication that the ostensible priority given to environmental justice at the federal level resulted in a change in performance of state regulatory agencies. That is, there is little indication of a response of state regulatory enforcement to the federal policy signal on environmental justice. State agencies did not systematically target firms in low-income and minority communities with additional enforcement actions in the years after the EPA stated its commitment to use its enforcement powers to pursue environmental justice goals (firms in Hispanic communities were an exception). Either the EPA did not push states to modify their enforcement strategies, or if it did, the agency was ignored.[12]

It is important to recognize that establishing consistent implementation throughout the entire U.S. environmental protection system is a difficult task. In addition, by no means is the challenge unique to environmental justice policy. Numerous reports over the years have concluded that the EPA needs to do a better job in its oversight of the implementation activities of its regional offices and state agencies (e.g., EPA Inspector General 1998, 2005, 2011; GAO 1996; 2000; 2007, 2012, 2014). Moreover, it is also must be recognized that lax EPA coordination and oversight does not necessarily mean that the EPA regions and states are neglecting to address environmental justice concerns. Although beyond the scope of this book, there are many examples of innovation on environmental equity issues in these agencies, particularly at the state level (Bonorris 2010; Ringquist and Clark 1999, 2002).

However, for every policy leader, there is also a policy laggard, and this is where national guidance and coordination is most critical. Too much variation across the country can erode confidence in the federal government, and it is particularly problematic with respect to environmental justice, which is predicated on the idea of *equal* protection. A real risk for the EPA is that consideration of environmental justice becomes ad hoc, varying from siting decision to siting decision, permit to permit, economic analysis to economic analysis, and so on. There is a fine line between instilling consistency in decision making and creating too much rigidity, especially in light of data and methodological limitations. However, without clear guidelines and institutional coordination across agencies, decision makers are left to invent and reinvent ways to incorporate environmental justice into their decisions, and as a consequence, there is extreme variation across decisions. This situation at best leaves an impression of unequal treatment, and at worst results in real unequal treatment.

A third factor that emerged throughout the book, if only implicitly, is that implementation of federal environmental justice policy has been stunted by a lack of steady leadership. Leadership on this issue at the EPA has been uneven over the past twenty years. The initial flurry of policy activity began during the tenure of Carol Browner in the Clinton administration,[13] but over the course of her eight-year term as EPA administrator, the issue waned as a focal point of agency activity. This is reflected, for instance, in the lack of policy guidance developed during this period, which slowed implementation of EO 12898 and other stated priorities of the agency considerably during the formative years of agency attention to the issue.

The eight years of the George W. Bush administration that followed are widely viewed as a period of general retrenchment in environmental policy (Vig 2013), and this seems to be the case specifically with environmental justice (Mohai, Pellow, and Timmons Roberts 2009). At the time, some in the environmental justice community worried that the EO 12898 itself would be rescinded. Although this did not happen, the EPA retreated in an important way from the goals articulated in EO 12898 by decoupling environmental justice from its historic focus on low-income and minority communities. Former EPA administrator Christine Todd Whitman issued a memorandum to agency leaders, reiterated four years later by her successor, Stephen Johnson, that on its face reaffirmed the EPA's commitment to environmental justice (Whitman 2001; Johnson 2005). However, the memorandum made a point of emphasizing that environmental justice is not only about addressing disproportionate risks for poor and minority groups. The EPA Inspector General noted in a 2004 report that the memo:

changed the focus of the environmental justice program by deemphasizing minority and low-income populations and emphasizing the concept of environmental justice for everyone. This action moved the Agency away from the basic tenet of the Executive Order and has contributed to the lack of consistency in the area of environmental justice integration." (EPA Office of the Inspector General 2004, p. 10)

A year later, the EPA's Office of Environmental Justice rearticulated this message, advising senior agency management that it

should recognize that the environmental justice program is not an affirmative action program or a set-aside program designed specifically to address the concerns of minority communities and/or low-income communities. To the contrary, environmental justice belongs to all Americans and it is the responsibility of Agency officials, as public servants, to serve all members of the public. (quoted in EPA Office of the Inspector General 2004, 10)

These policy statements sent strong signals from the top of the agency that the EPA was not going devote much effort to pursuing environmental justice.

As has been documented throughout the book, environmental justice returned to the top of the EPA's agenda during the first term of the Obama administration. President Obama's first EPA administrator, Lisa Jackson, showed a genuine passion for the issue. Under Jackson's leadership, the EPA began an enormous policy project, known as *Plan EJ 2014*, designed to bring equity considerations to the center of agency decision making in a way that was perhaps envisioned by EO 12898. As described next in more depth, *Plan EJ 2014* includes the development of detailed policy guidance and strategies to pursue equity goals in rulemaking, permitting, enforcement, community-based programs, and other programs throughout federal agencies. In addition, the project has initiated the development of new tools for the agency and external stakeholders in the area of science, law, information, and resources. Collectively, the plan reflects a concerted and comprehensive effort to integrate environmental justice into the day-to-day activities of the EPA.

Policy priorities come and go in all federal agencies, so the ebb and flow of attention to environmental justice over a two-decade period should come as no surprise. Presidents and cabinet officials have enormous discretion as to where to devote their agencies' time, energy, and human and financial resources, and the agenda is crowded by many important issues. The significance of how they use their discretion, however, is magnified when the actions to be implemented stem from executive orders, strategy documents, or other policy plans. These purely administrative measures lack the enforcement mechanisms embedded in legislation. Most health, safety, and environmental statutes in particular contain built-in regulatory deadlines and mandates backed by liberal public standing provisions that enable citizens and interest groups to hold an agency's "feet to the fire." Moreover, congressional programmatic and budgetary oversight can keep an agency further focused on its legislative commitments. Without these mechanisms, there simply has been nothing to force the federal government to follow through on its stated environmental justice commitments.

The consequences of inconsistent agency leadership have been substantial. The development of guidance has been slow, and there has been ineffective policy coordination across EPA regions and the states. The EPA has also been reticent to use its discretion to more aggressively pursue environmental justice goals in areas where it has the capacity to act. This issue was highlighted throughout the book. As just a few examples, the agency has not regularly taken into account equity in setting new standards; it has not as a matter of practice insisted on the evaluation of the distributional impacts of new rules in economic analysis; it has not routinely reached out to involve poor and minority communities in decision making; it has not systematically targeted regulatory enforcement actions in low-income and minority communities; and it has not aggressively resolved allegations of Title VI violations. In addition, there is little evidence that the EPA has used discretion identified by its own legal counsel and others on incorporating environmental justice into permitting under major environmental statutes. There has been a trend toward more consideration of environmental justice in these areas in recent years, which given the emphasis that the Obama administration has placed on pursuing equity goals, shows the importance of leadership.

Looking to the Future

The critical question, moving forward, is whether the recent policy developments represent just a temporary shift in priorities or whether they mark the beginning of a sustained commitment to fulfill the promises of federal environmental justice policy. If the past twenty years have largely been a disappointment, what should we expect in the years and decades to come?

As this book was being conceptualized, the EPA was embarking on an ambitious new initiative—*Plan EJ 2014*—to, in essence, restart its strategy to integrate equity considerations into its decision making. Released in September 2011, *Plan EJ 2014* provided, in the words of former EPA

Administrator Jackson, "a road map that will enable us to better integrate environmental justice and civil rights in our programs, policies, and daily work" (EPA 2011).

Plan EJ 2014 is an agencywide initiative that extends to the EPA's core regulatory activities, as well as to its research, legal, civil rights, and community outreach programs. There are several noteworthy features of the initiative. First, the EPA has developed new policy and technical guidance in the areas of permitting, rulemaking, and enforcement. With respect to permitting, the agency has focused on ensuring that communities overburdened by environmental risks are involved in a meaningful way in its permitting decisions. Specifically, each EPA regional office has developed plans that describe how and when they will engage with communities potentially affected by the granting of a new permit, and issued a best-practices document targeted at permit applicants to raise their sensitivity to equity issues. In the area of rulemaking, the agency followed up on the issuance of its interim guidance in July 2010 (referenced previously), with draft technical guidance on how to analytically incorporate environmental justice issues in the development of new standards (along the lines discussed in chapter 5; also see EPA 2013). The technical guidelines provide a procedural framework, with the intention of standardizing how the agency handles equity considerations when setting standards as part of the statutes it implements. In terms of enforcement, *Plan EJ 2014* points to a number of activities that the agency has pursued to more fully incorporate equity considerations into enforcement priority setting and decisions. Among the activities highlighted include new internal reporting and tracking requirements, a policy to integrate environmental justice concerns into assessments for criminal investigations, making equity a priority in RCRA enforcement, and emphasizing equity in NEPA decisions (EPA 2011).

In addition to the development of this new guidance, *Plan EJ 2014* consists of initiatives in several other areas of EPA programs and activities. The strategy includes efforts to strengthen community-based programs and capacity in low-income and minority communities, to promote action on environmental justice throughout the executive branch of government, to develop new tools in the areas of science and research, to provide legal assistance to agency decision makers on how to better use their existing administrative discretion under environmental statutes, and to develop informational tools and other resources to enhance internal decision making and the ability of communities to interface with the agency (EPA 2012). Finally, *Plan EJ 2014* seeks to build on existing programs in areas such as urban water quality and pesticide worker safety, as well as improve on the EPA's handling of cases in its enforcement of Title VI of the Civil Rights Act of 1964. Without question, *Plan EJ 2014* represents both an important statement about the importance of environmental justice and a comprehensive strategy that suggests a clear intention to prioritize it throughout the activities of the EPA. In this respect, the EPA has begun to put the goals and aspirations of EO 12898 into practice—albeit twenty years later.

At the time of this writing, the development of *Plan EJ 2014* was ongoing, and it will certainly take some time to evaluate its effects on environmental decision making at the EPA. But even at this preliminary stage, there are some lessons to be learned from the analyses in this book. Clearly, one reason for optimism is that the EPA is finally developing the types of policy and technical guidance documents that have been sorely lacking in the previous attempts to implement EO 12898 and other federal environmental justice reforms. This guidance could go a long way toward ensuring that environmental justice considerations are given regular and consistent treatment by providing permit writers, rulemakers, and enforcement officials with the information and tools they need to assess, and if necessary, take into account equity when making their decisions. In addition, the new strategy calls for routine and deep engagement with stakeholders directly affected by agency decisions, particularly vulnerable segments of the population, as well as a commitment to pursue equity through the use of Title VI of the federal Civil Rights Act and NEPA. In this respect, many of the shortcomings of implementation efforts documented in this book are recognized.

Another distinguishing feature of *Plan EJ 2014* is that it clearly has, not just the endorsement of EPA leadership, but its imprint. The development of the strategy was led and coordinated directly out of former administrator Jackson's office, which sent a clear signal through the organization that it was an important policy matter and an important political priority. Moreover, there was a deliberate effort to get "buy-in" by integrating staff in offices throughout the agency in its development, including media-specific programs, the functional offices (e.g., enforcement, policy, research, and legal), as well as the regional offices. The involvement of the full agency should foster improved policy coordination of environmental justice goals as *Plan EJ 2014* transitions from development into implementation. It should be reiterated, however, that much of the implementation of federal pollution control laws is handled at the state level, and *Plan EJ 2014* is nearly silent on how the EPA will see to it that these new priorities are integrated into the decisions made by state agencies.

Despite these notable aspects of *Plan EJ 2014*, there is also reason to be cautious about its ultimate impact on environmental decision making. Here again, we can point to lessons learned from the analyses in this book. First, it is important to recognize that *Plan EJ 2014* is a declaration of agency policy. As a strategy document, it is very comprehensive—certainly more so than any previous strategy on environmental justice generated by the EPA (or any other federal agency, for that matter). But it is just that, and only that. Still to be determined is how the various new guidance documents and other new practices and programs will be used once they are finalized. In other words, the real test is in how the elements of *Plan EJ 2014* are implemented in future environmental decision making. Will permit writers working in EPA regional offices follow through on their plans to actively engage with residents in low-income and minority communities when making decisions? Will new rules be promulgated and analyzed in a way that fully considers their distributional effects on vulnerable populations? Will the EPA fulfill its new commitment to pursue complaints of civil rights violations under Title VI? Are future regulatory enforcement strategies informed by considerations of their environmental justice implications? These are just a few of the questions that can only be answered in the years to come.

More generally, there should be a concern about the sustainability of the new initiative over the long run. *Plan EJ 2014* is neither a legislatively mandated regulation,[14] nor even an executive order. It is a statement of agency policy and strategy that could be maintained in full, kept in part, or withdrawn (or just ignored) completely by a future presidential administration. In this regard, it is similar to other unilateral administrative actions, and there is simply no guarantee that a future president or EPA would devote the same resources or attention to environmental justice as has been seen in recent years. As discussed previously, over the course of the now-twenty years since EO 12898 was first signed, there has been significant ebb and flow of attention to this issue. Whether we are witnessing a short-term apex or the beginning of a sustained commitment to address the ongoing problem of environmental inequities remains to be seen.

This is an important question for policymakers, academic scholars, and, most important, the residents of low-income and minority communities that continue to be disproportionately burdened by environmental risks. Much of the past thirty years has been spent documenting the presence and degree of environmental inequities. To be sure, this is a worthy endeavor, and there remain important outstanding questions that require further investigation. However, it is just as important to devise an effective policy response and to evaluate the actions taken by government to addresses persistent violations of environmental justice. While the analyses in this book suggest a disappointing record to date for the federal government (and particularly the EPA), there are also reasons to be optimistic about the future. Environmental inequities present complicated political, economic, and social questions for government. The task of minimizing, let alone eliminating, these inequities is a challenging one that may never be fully realized. But a government that is fully attentive to and thoughtful about these inequities in its decision making will be one that comes closer to fulfilling its promise to deliver environmental justice.

Notes

1. To date (at least to my knowledge), there has not been a large-scale empirical analysis of EPA permit decisions to evaluate the degree to which equity considerations have been incorporated into agency decision making.

2. Of course, permit cases that come before the EAB reflect a small number which have been challenged on some grounds, environmental justice or otherwise.

3. This is obviously a complicated legal question, subject to differences of opinion in the interpretation of various statutes. In the lead paragraph of the memo, Gary Guyz, then the general counsel of the EPA, clearly noted as much: "Although the memorandum presents interpretations of EPA's statutory authority and regulations that we believe are legally permissible, it does not suggest that such actions would be uniformly practical or feasible given policy or resource considerations or that there are not important considerations of legal risk that would need to be evaluated" (EPA Office of General Counsel 2000, p. 1). The general point here is that the EPA has not employed its permitting powers or procedures to achieve environmental justice goals in any serious way.

4. This conclusion has been reached by others as well, most notably Banzhaf (2011), the Government Accountability Office (GAO 2005), and Vajjhalla, Van Epps, and Szambelan (2008).

5. The analysis searches for average effects, so it is certainly possible (indeed likely) that regulatory enforcement officials were motivated to pursue some firms in low-income and minority communities, in part to achieve environmental justice goals.

6. The analysis in chapter 7 focused on the CAA, but in a separate analysis, I reached similar conclusions with respect to the CWA and the RCRA (Konisky 2009).

7. The claims repeatedly made in the literature almost always refer to a study by Lavelle and Coyle (1992), which concluded that there was such a bias in penalty outcomes. As discussed in both chapters 7 and 8, subsequent work has shown that the findings in this study were spurious and overstated.

8. Again, this is not to suggest that there is no bias in some specific cases. The analyses reviewed by Gross and Stretesky are looking for general trends across many cases, and the results they summarize reflect the average effects.

9. The EPA did release an environmental justice strategy in 1995 and an implementation plan a year later, but these were written at a high level, without the detail needed to guide decision making in any specific way.

10. Of course, the fact that guidance was often written in a general way was deliberate, so as not to constrain decision making with rigid guidelines, or perhaps to create false expectations that decisions would be markedly different.

11. According to one recent accounting, as of November 2010, 96 percent of the programs that could be delegated to the states have been delegated (Environmental Council of the States 2013). As a concrete example of the implications of this delegation, state administrative agencies carry out more than 95 percent of all enforcement actions under laws such as the CAA, the CWA, and the RCRA (Environmental Council of the States 2006).

12. Of course, there is significant variation across the states, and EPA coordination may have been more effective in some states than others. One mechanism that has been used in some cases is EPA-State Performance Partnership Agreements, which are cooperative agreements to address and align priorities with the states.

13. As was discussed in chapter 2, William Reilly, administrator of the EPA during the George H.W. Bush administration, did begin paying federal attention to environmental justice, especially with the creation of a working group on environmental equity that issued the agency's first report on the subject in 1992 (EPA 1992).

14. The EPA emphasizes this point. As an example, the agency clearly states on its environmental justice website: "Plan EJ 2014 is **not a rule or regulation**. It is a *strategy* to help integrate environmental justice into EPA's day to day activities" (emphasis in original). See http://www.epa.gov/environmentaljustice/plan-ej/index.html.

References

Banzhaf, H. Spencer. 2011. Regulatory Impact Analyses of Environmental Justice Effects. *Journal of Land Use & Environmental Law* 27 (1): 1–30.

Bell, Michelle L., and Keita Ebisu. 2012. Environmental Inequality in Exposures to Airborne Particulate Matter Components in the United States. *Environmental Health Perspectives* 120 (12): 1699–1704.

Bonorris, Steven. 2010. *Environmental Justice for All: A Fifty-State Survey of Legislation, Policies, and Cases.* 4th ed., American Bar Association and Hastings College of the Law. http://gov.uchastings.edu/public-law/docs/ejreport-fourth edition1.pdf.

Bullard, Robert D., Paul Mohai, Robin Saha, and Beverly Wright. 2007. *Toxic Wastes and Race at Twenty, 1987–2007.* Cleveland, OH: United Church of Christ, Justice and Witness Ministries.

Clark, Lara P., Dylan B. Millet, and Julian D. Marshall. 2014. National Patterns in Environmental Injustice and Inequality: Outdoor NO_2 Air Pollution in the United States. *PLoS ONE* 9 (4): e94431. doi:.10.1371/journal.pone.0094431

Cohen, Mark A. 1999. Monitoring and Enforcement of Environmental Policy. In *International Yearbook of Environmental and Resource Economics 1999/2000,* ed. Henk Folmer and Tom Tietenberg, 44–106. Northampton, MA: Edward Elgar.

Environmental Council of the States. 2006. State Environmental Agency Contributions to Enforcement and Compliance, 2000-2003. Washington, DC: Environmental Council of the States.

Environmental Council of the States. 2013. Delegation by Environmental Act. http://www.ecos.org/section/states/enviro_actlist.

Environmental Protection Agency (EPA). 1992. *Environmental Equity: Reducing Risk for All Communities.* Vol. 1. Workgroup Report to The Administrator. EPA230-R-92-008. http://www.epa.gov/environmentaljustice/resources/reports/annual-project-reports/reducing_risk_com_vol1.pdf.

Environmental Protection Agency (EPA). 1996. *1996 Environmental Justice Implementation Plan.* Washington DC. EPA/300-R-96-004. April.

Environmental Protection Agency (EPA). 2004. *Toolkit for Assessing Potential Allegations of Environmental Injustice.* EPA 300-R-04-002. November.

Environmental Protection Agency (EPA). 2010. *EPA's Action Development Process: Interim Guidance on Considering Environmental Justice during the Development of an Action.* July. http://www.epa.gov/environmentaljustice/resources/policy/considering-ej-in-rulemaking-guide-07-2010.pdf

Environmental Protection Agency (EPA). 2011. *Plan EJ 2014.* September 2011. http://www.epa.gov/environmentaljustice/resources/policy/plan-ej-2014/plan-ej-2011-09.pdf.

Environmental Protection Agency (EPA). 2013. *Draft Technical Guidance for Assessing Environmental Justice in Regulatory Analysis.* August. http://yosemite

.epa.gov/sab/sabproduct.nsf/0/0F7D1A0D7D15001B8525783000673AC3/$File/ EPA-HQ-OA-2013-0320-0002[1].pdf.

Environmental Protection Agency (EPA), Office of Inspector General. 1998. *Consolidated Report on OECA's Oversight of Regional and State Air Enforcement Programs.* E1GAE7–03–0045–8100244. Washington, DC.

Environmental Protection Agency (EPA), Office of Inspector General. 2004. Memorandum from Gary S. Guzy to Steve A. Herman, Robert Perciasepe, Timothy Fields, Jr., and J. Charles Fox, "EPA Statutory and Regulatory Authorities Under Which Environmental Justice Issues May Be Addressed in Permitting," December 1. http://www.epa.gov/environmentaljustice/resources/policy/ej_permitting _authorities_memo_120100.pdf.

Environmental Protection Agency (EPA), Office of Inspector General. 2004. *EPA Needs to Consistently Implement the Intent of the Executive Order on Environmental Justice.* Report No. 2004-P-00007.

Environmental Protection Agency (EPA), Office of Inspector General. 2005. *Efforts to Manage Backlog Water Discharge Permits Need to Be Accompanied by Greater Program Integration.* Report No. 2005-P-00018. Washington, DC.

Environmental Protection Agency (EPA), Office of the Inspector General. 2011. *EPA Must Improve Oversight of State Enforcement.* Report No. 12-P-0113. December 9. Washington, DC.

Faber, Daniel R. 2008. The Proposed Brockton Power Plant: Environmental Disparities in Brockton, MA. March 27. http://www.stopthepower.net/images/Brockton _Power_Plant_Report.pdf.

General Accounting Office (GAO). 1996. *Water Pollution: Differences among the States in Issuing Permits Limiting the Discharge of Pollutants.* GAO/RCED-96-42. Washington, DC.

General Accounting Office (GAO). 2000. *Environmental Protection: More Consistency Needed Among EPA Regions in Approach to Enforcement.* GAO/RCED-00-108. Washington, D.C.

Government Accountability Office (GAO). 2005. *EPA Should Devote More Attention to Environmental Justice When Developing Clean Air Rules.* GAO-05-289. July.

General Accountability Office (GAO). 2007. Environmental Protection: EPA-State Enforcement Partnership Has Improved, but EPA's Oversight Needs Further Enhancement. GAO-07-883. July 31. Washington, DC.

General Accountability Office (GAO). 2012. Nonpoint Source Water Pollution: Greater Oversight and Additional Data Needed for Key EPA Water Program. GAO-12-335. January 13. Washington, DC.

General Accountability Office (GAO). 2014. Clean Water Act: Changes Needed if Key EPA Program Is to Help Fulfill the Nation's Water Quality Goals. GAO-14-80. January 13. Washington, DC.

Gray, Wayne B., and Jay P. Shimshack. 2011. The Effectiveness of Environmental Monitoring and Enforcement: A Review of the Empirical Evidence. *Review of Environmental Economics and Policy* 5 (1): 3–24.

Jackson, Lisa P. 2010. Seven Priorities for EPA's Future. Memorandum to All EPA Employees. January 12, 2010. http://blog.epa.gov/administrator/2010/01/12/seven-priorities-for-epas-future/ (last accessed August 20, 2013).

Johnson, Stephen L. 2005. Reaffirming the U.S. Environmental Protection Agency's Commitment to Environmental Justice. Memorandum to Assistant Administrators et al., November 4, 2005. http://www.epa.gov/region7/ej/pdf/admin-ej-commit-letter.pdf.

Klyza, Christopher McGrory, and David Sousa. 2008. *American Environmental Policy, 1990–2006: Beyond Gridlock*. Cambridge, MA: MIT Press.

Konisky, David M. 2009. The Limited Effects of Federal Environmental Justice Policy on State Enforcement. *Policy Studies Journal* 37 (3): 475–496.

Lavelle, Marianne, and Marcia Coyle. 1992. Unequal Protection: The Racial Divide in Environmental Law. *National Law Journal* (September): 21.

Miranda, Marie Lynn, Sharon E. Edwards, Martha H. Keating, and Christopher J. Paul. 2011. Making the Environmental Justice Grade: The Relative Burden of Air Pollution Exposure in the United States. *International Journal of Environmental Research and Public Health* 8 (6): 1755–1771.

Mohai, Paul, David Pellow, and J. Timmons Roberts. 2009. Environmental Justice. *Annual Review of Environment and Resources* 34: 405–430.

National Academy of Public Administration. 2001. Environmental Justice in EPA Permitting: Reducing Pollution in High-Risk Communities is Integral to the Agency's Mission. December. http://www.epa.gov/environmentaljustice/resources/reports/annual-project-reports/napa-epa-permitting.pdf.

National Environmental Justice Advisory Council (NEJAC). 2010. *National Environmental Justice Advisory Council Meeting*, January 27–29, 2010. Meeting transcript from January 28, 2010. http://www.epa.gov/compliance/ej/resources/publications/nejac/nejacmtg/nejac-meeting-trans-012810.pdf (last accessed August 20, 2013).

Ringquist, Evan J., and David H. Clark. 1999. Local Risks, States' Rights, and Federal Mandates: Remedying Environmental Inequities in the U.S. Federal System. *Publius: The Journal of Federalism* 29 (2): 73–93.

Ringquist, Evan J., and David H. Clark. 2002. Issue Definition and the Politics of State Environmental Justice Policy Adoption. *International Journal of Public Administration* 25 (2&3): 351–389.

Vajjhala, Shalini P., Amanda Van Epps, and Sarah Szambelan. 2008. *Integrating EJ into Federal Policies and Programs: Examining the Role of Regulatory Impact Analyses and Environmental Impact Statements*. RFF Discussion Paper 08-45. Washington, DC: Resources for the Future.

Vig, Norman J. 2013. Presidential Powers and Environmental Policy. In *Environmental Policy: New Directions for the 21st Century*, eds. Norman J. Vig and Michael E. Kraft, 84–108. Washington, DC: CQ Press.

Whitman, Christine Todd. 2001. EPA's Commitment to Environmental Justice. Memorandum to Assistant Administrators et al., August 9, 2001. http://yosemite .epa.gov/opa/admpress.nsf/89745a330d4ef8b9852572a000651fe1.

Wilson, Adrian. 2012. *Coal Blooded: Putting Profits before People.* Prepared for the National Association for the Advancement of Colored People, the Indigenous Environmental Network, and the Little Village Environmental Justice Organization. http://naacp.3cdn.net/afe739fe212e246f76_i8m6yek0x.pdf (last accessed August 20, 2013).

10
Contributors

Dorothy M. Daley is an associate professor at the University of Kansas, with a joint appointment in the School of Public Affairs and Administration and the Environmental Studies Program. Her research interests are in the areas of environmental and public health policy, state politics and federalism, and urban redevelopment. She has published extensively on the role of public participation in environmental decision making and hazardous waste policy implementation. Daley holds a Ph.D. in ecology, with a specialization in environmental policy analysis from the University of California, Davis.

Eileen Gauna is a professor of law at the University of New Mexico School of Law, where she teaches courses in environmental law, environmental justice, energy law, climate law, administrative law, and property. Her publications include *Environmental Justice: Law, Policy, and Regulation* (with Clifford Rechtschaffen and Catherine O'Neil), a leading textbook on environmental justice, along with numerous reports, chapters, and law review articles on environmental law and environmental justice. Professor Gauna is a member scholar of the Center for Progressive Reform and the American Law Institute; she also serves on the ABA Task Force on Sustainable Development.

Elizabeth Gross is an attorney currently practicing in Edwards, Colorado, primarily in the area of real estate law. She also has practiced law in the local government and municipal realms, including performing research into a variety of areas of constitutional law. In addition, she has served as a prosecutor in multiple jurisdictions. She is a graduate of Emory University and Cornell Law School, where she served as an editor of the *Cornell International Law Journal*.

David M. Konisky is an associate professor at the McCourt School of Public Policy at Georgetown University. Konisky's research focuses on U.S. politics and public policy, with particular emphasis on regulation, environmental politics and policy, state politics, and public opinion. His work has been published in leading political science and public policy journals, and his previous research on environmental justice has been funded by the Russell Sage Foundation. Konisky is also the author of two books, *Cheap and Clean: How Americans Think about Energy in the Age of Global Warming* (MIT Press, 2014, with Steve Ansolabehere) and *Superfund's Future: What Will It Cost?* (RFF Press, 2001, with Kate Probst).

Konisky earned his Ph.D. in political science from the Massachusetts Institute of Technology.

Douglas S. Noonan is an associate professor at the School of Public and Environmental Affairs at Indiana University-Purdue University Indianapolis and the director of research at the Indiana University Public Policy Institute. His research focuses on a variety of policy and economics issues related to the urban environment, neighborhood dynamics, and quality of life. His research has been sponsored by many organizations (e.g., the National Science Foundation, Environmental Protection Agency, Lincoln Institute of Land Policy, and National Endowment for the Arts) on topics like environmental justice, air quality, green urban revitalizations, and cultural economics. Noonan earned his Ph.D. in public policy at the University of Chicago.

Tony G. Reames is a postdoctoral research fellow at the School of Natural Resources and Environment at the University of Michigan. His research interests include urban governance, environmental justice, climate change, and energy policy. Reames holds a Ph.D. in public administration from the University of Kansas and a master's degree in engineering management and a bachelor's degree in civil engineering from Kansas State and North Carolina A&T State, respectively.

Christopher Reenock is an associate professor in the Department of Political Science of the College of Social Science and Public Policy at Florida State University. Professor Reenock's research in public policy focuses on the consequences of agency design choices on policy implementation and equity. His work has appeared in a variety of outlets, including the *American Journal of Political Science*, *Journal of Politics*, *Legislative Studies Quarterly*, and *Journal of Public Administration Research and Theory*. Reenock earned his Ph.D. in political science from Pennsylvania State University.

Ronald J. Shadbegian is an economist at the Environmental Protection Agency's National Center for Environmental Economics and an adjunct professor of economics in the Economics Department and at the McCourt School of Public Policy at Georgetown University. He served as senior economist for energy and the environment at the Council of Economic Advisers from 2013–2014. His research focuses on the costs of complying with environmental regulations and the economic impacts of environmental regulations on environmental justice, employment, productivity, investment, environmental performance, and technological change. He received his Ph.D. in economics at Clark University in 1991.

Paul Stretesky is a professor of criminology in the Department of Social Sciences and Languages at Northumbria University in Newcastle, England. He is co-editor of the Ashgate series *Green Criminology*. His recent books include *Exploring Green Criminology: Toward a Green Criminological Revolution* (2014, with M. J. Lynch); *Treadmill of Crime: Political Economy and Green Criminology* (2013, with M. Long and M. J. Lynch); *Environmental Crime, Law, and Justice* (2008, with R. G. Burns and M. J. Lynch); *Guns, Violence, and Criminal Behavior: The Offender's Perspective* (2009, with M. R. Pogrebin and N. P. Unnithan) and *Radical Criminology* (2011, with M. J. Lynch). His research is focused on issues of environmental justice and crime. He is also engaged in community-based research for the nonprofit organization Families of Homicide Victims and Missing Persons.

Ann Wolverton has worked as an economist at the National Center for Environmental Economics in the Environmental Protection Agency since 2001. Her key areas of research include environmental justice, climate change, the effectiveness of voluntary programs, and market incentives. She served as the senior economist for environmental and natural resources at the Council of Economic Advisers in 2006–2007 and again in 2009–2010. Prior to working at the EPA, Ann was a senior associate at ICF Consulting. She holds a Ph.D. in economics from the University of Texas at Austin and bachelor's degrees in English and economics from Arizona State University.

Index

American and Comparative Environmental Policy

Sheldon Kamieniecki and Michael E. Kraft, series editors

Russell J. Dalton, Paula Garb, Nicholas P. Lovrich, John C. Pierce, and John M. Whiteley, *Critical Masses: Citizens, Nuclear Weapons Production, and Environmental Destruction in the United States and Russia*

Daniel A. Mazmanian and Michael E. Kraft, editors: *Toward Sustainable Communities: Transition and Transformations in Environmental Policy*

Elizabeth R. DeSombre, *Domestic Sources of International Environmental Policy: Industry, Environmentalists, and U.S. Power*

Kate O'Neill, *Waste Trading among Rich Nations: Building a New Theory of Environmental Regulation*

Joachim Blatter and Helen Ingram, editors, *Reflections on Water: New Approaches to Transboundary Conflicts and Cooperation*

Paul F. Steinberg, *Environmental Leadership in Developing Countries: Transnational Relations and Biodiversity Policy in Costa Rica and Bolivia*

Uday Desai, editor, *Environmental Politics and Policy in Industrialized Countries*

Kent Portney, *Taking Sustainable Cities Seriously: Economic Development, the Environment, and Quality of Life in American Cities*

Edward P. Weber, *Bringing Society Back In: Grassroots Ecosystem Management, Accountability, and Sustainable Communities*

Norman J. Vig and Michael G. Faure, editors, *Green Giants? Environmental Policies of the United States and the European Union*

William Ascher, Toddi Steelman, and Robert Healy, *Knowledge in the Environmental Policy Process: Re-Imagining the Boundaries of Science and Politics*

Michael E. Kraft, Mark Stephan, and Troy D. Abel, *Coming Clean: Information Disclosure and Environmental Performance*

Paul F. Steinberg and Stacy D. VanDeveer, editors, *Comparative Environmental Politics: Theory, Practice, and Prospects*

Judith A. Layzer, *Open for Business: Conservatives' Opposition to Environmental Regulation*

Kent Portney, *Taking Sustainable Cities Seriously: Economic Development, the Environment, and Quality of Life in American Cities,* second edition

Raul Lejano, Mrill Ingram, and Helen Ingram, *The Power of Narrative in Environmental Networks*

Christopher McGrory Klyza and David J. Sousa, *American Environmental Policy: Beyond Gridlock,* updated and expanded edition

Andreas Duit, editor, *State and Environment: The Comparative Study of Environmental Governance*

Joseph F. C. DiMento and Pamela Doughman, editors, *Climate Change: What It Means for Us, Our Children, and Our Grandchildren,* second edition

David M. Konisky, editor, *Failed Promises: Evaluating the Federal Government's Response to Environmental Justice*